CLONING
Miracle or Menace?

CLONING

MIRACLE OR MENACE?

Lane P. Lester, Ph.D.
with
James C. Hefley

Tyndale House
Publishers, Inc.
Wheaton, Illinois

This book is
gratefully dedicated to
DR. DUANE T. GISH
whom God used
to show me the
total trustworthiness of
his Holy Word.

CONTENTS

1

A "BRAVE NEW WORLD"

They said it was impossible.

A motorized buggy which travels at the incredible speed of thirty miles an hour.

A mechanical "bird" that flies across oceans.

A "gun" that can shoot a missile ten thousand miles, hit a yard-wide target, and kill five million people.

Men on the moon. A space vehicle able to pierce the rings of Saturn 900 million miles from earth.

And now the cloning of a man in his exact genetic likeness, producing an offspring that is both his son and his twin.

I hear it said this, too, is impossible. One Christian leader has said, "If it happens, my faith is destroyed."

David Rorvik, an award-winning science writer who formerly worked for *Time,* claims a man has already been cloned in secret. Rorvik's book, appropriately called *In His Image,* describes how an eccentric multimillionaire paid to have himself cloned by a master scientist. The respectable J. B. Lippincott Company that published *In His Image,* said in its foreword, "The account that follows is an astonishing one. The author assures us it is true. We do not know."[1]

Rorvik's technical information appears quite accurate, but he uses fictitious names for his principal characters, and he offers no external evidence of the actual cloning. On the basis of the given information, I believe his story is a clever fake.

However, on a scale of one to ten, I'd say that research biologists are now at eight and approaching nine on their way to cloning a human. Barring worldwide disaster or unexpected governmental crackdowns on research, most scientists predict a human will be cloned in ten to twenty years.

"Clone" is a collective word, as is "gang" or "fraternity," in that it refers to one or more individuals. Cloning is not a new idea. Horticulturists have been cloning for years to produce new plants from cuttings, rather than from seeds. Frogs were cloned over a decade ago and research on mice is now going forward. Clones of pigs, sheep, cows, and monkeys, as well as humans, are on the horizon.

Scientists propose to take a cell from a human and "trick" it into multiplying until it forms a human embryo, completely bypassing the normal process of conception. This embryo will then be implanted in a woman's uterus where it will presumably develop into a human fetus and be born normally. But this child will be the exact genetic twin of its parent. A reputable scientist, Dr. Landrum B. Shettles, says he reached the embryo stage three times in cloning a human, then decided to end the experiment.

All the hundred trillion or so cells in your body (except red blood cells which lose their nuclei during development) contain your total genetic blueprint, right down to the shape of your ears, the color of your eyes, and the loops and whorls of your fingerprints. Theoretically, you could be cloned billions of times if there were enough wombs, human or artificial, to carry your offspring to term. However, for very good reasons which I'll explain later, an ordinary body cell will probably not be used in the first clonings, but rather one of the special reproductive cells which ordinarily produce sperm in males or eggs in the case of females.

Geniuses and other persons with special gifts might even be cloned after death. Cells live for hours after vital organs have expired. It would be no trick to freeze a body to preserve enough cells for cloning a select person. A wealthy person might leave some of his cells in a laboratory culture and bequeath a grant to the lab for his cloning. However, I expect the first human cloned to be from a live donor.

Dr. James Watson, a Nobel prize winner, thinks the initial response will be "one of despair. The concept of the child-parent bond, not to mention everyone's values about the individual's uniqueness, could be changed beyond recognition,"[2] he says. The cloning of a human will not destroy my faith in God. The event will be cataclysmic and some weak believers may flip and say, "This proves there is no God." But no one can really predict what the effect on our culture will be.

There's no doubt that human cloning will unleash a torrent of questions that strike at the roots of basic values cherished in our society.

What will happen to our concept of uniqueness and individuality? Will members of a clone suffer identity crises?

Will cloning destroy the two-parent family?

Will a clonal child be made in man's image or God's likeness? Will he be an eternal soul? Fully human? Will he be responsible for his sins, or can he blame his parent?

The response of the Christian community may be critical. Will we condemn cloning as a mockery of God's will for mankind? How will we oppose cloning?

How will cloning be regarded by society at large? The first cloning of a man will certainly be a media event of unprecedented proportion. We'll be inundated by television specials, movies, books, and feature articles. I expect that legislation will be introduced in many countries to forbid further cloning of humans.

Vigorous debate is already taking place among scientists, ethicists, philosophers, and theologians. Dr. Watson argues that human cloning could cause the collapse of western civilization.

At the other extreme is Joseph Fletcher, best known for his doctrine of "situation ethics" which challenges biblical morality. Fletcher endorses cloning when "the greatest good for the greatest number is served."[3]

Scientists were talking seriously about human cloning for years before the public became interested. An important preliminary step toward human cloning was the announcement on July 25, 1978, by two English doctors, of the birth of the world's first "test-tube" baby. Millions of people were fascinated and wanted to know how a baby could be born in a test-tube. Was this a prelude to cloning? They discovered that the birth didn't quite live up to the headline, but that the achievement did bring human cloning one step closer.

I'll discuss the so-called "test-tube" baby in greater depth in chapter six. Briefly, this is what happened. Twenty-nine-year-old Lesley Brown had blocked fallopian tubes, a problem commonly reported by gynecologists, which prevents an otherwise fertile woman from achieving conception by sexual intercourse. Like many other women who find themselves unable to conceive, Mrs. Brown was referred by her doctor to a specialist. "Don't worry," Dr. Patrick Steptoe assured her, "you're just the right age for my purpose."

Medical researchers had been working on the procedure for several years. In 1974 Dr. Douglas Bevis told the British Medical Association that he had achieved the birth of three babies by *in vitro* fertilization (IVF, meaning "in glass"—outside the human body and in an artificial environment). When he declined to identify them and give specific details, his story was disbelieved. However, Steptoe and his partner, Dr. Robert Edwards, spared nothing and produced the Brown baby in connection with a lucrative publicity contract the Browns had signed with the London *Daily Mail.*

In the epochal achievement, the doctors surgically removed an egg cell from Mrs. Brown and placed it in a nutrient solution along with a sample of her husband's sperm. One of the sperm cells entered the egg cell and conception occurred.

The fertilized egg *(zygote)* divided and grew into an embryo which the researchers successfully implanted into Mrs. Brown's womb. A little less than nine months later a healthy baby was born.

A second IVF baby was born the following October 3 in Calcutta, India. By 1985 there will probably be hundreds, if not thousands of children who were conceived and implanted by the Steptoe-Edwards method.

Their feat solves an intermediate problem in cloning. How do you nurture a developing embryo and implant it in a woman's womb? Scientists are now working hard on the initial step, "persuading" a one-parent human cell to grow into an embryo in the lab. Once this is accomplished, the embryo can be implanted in a woman's womb, as was done for Baby Louise Brown, where it will presumably develop and be born as other humans are.

Apart from cloning, the process by which the "test-tube" baby was born immediately makes possible some startling new scenarios in motherhood. A "natural" mother could provide her egg for fertilization. A "bearing" mother could receive the implant and carry the child to term. Or to reverse the situation, another man's sperm could have been used to fertilize Mrs. Brown's egg. Then she or another woman could have given birth. Already we're hearing of women volunteering to bear the child of another. One woman in London has volunteered to bear her sister's child by way of an embryo implant, because the sister is unable to give birth. Another woman has offered to "rent" her womb for this purpose.

Can you imagine the moral and legal questions that will arise from such tangles?

Go a step further. Joan Collins can provide a fertile egg, but she does not wish to go to the trouble of bearing a child. Mary Hendrix agrees to bear her child for $10,000. Joan and her husband Harry sign a contract with Mary to this effect. Harry, however, is sterile because of a childhood case of mumps. No matter. The doctor arranges for a sperm donor. Who will be the "real" mother and "real" father of this child?

Artificial insemination by donor can provide us with some insight into what is to come. It's been around for years. Because of the secrecy involved and the question of parenthood when donor sperm is used, nobody knows how many children produced by AID (artificial insemination by donor) are alive. Estimates in the United States range from one hundred thousand to five hundred thousand.

AID simply enables a female to get pregnant apart from direct sexual intercourse. Around half of all American dairy cattle are now conceived this way.

Sperm banks are used widely in scientific animal husbandry. One bull may "father" a thousand calves in one year. The semen is kept frozen in sperm banks until ready for insemination.

The history of science indicates that daring new experimentations are first done in animals, then in people. The use of frozen sperm is a prime example of this general rule of biological science. Through the same technology involved in animal insemination, fifteen human sperm banks are now operating in the United States, although many doctors prefer fresh semen given less than a half hour before insemination.

Chapter seven will tell you all you want to know about

14

AID. Beyond AID and IVF lie "parent stores" where an aspiring couple or single woman can start a family by picking up a frozen embryo (perhaps produced on a fertility "farm," perhaps the result of cloning) which is ready for implantation. This may be classified as adoption before birth. Based on the way moral values are declining in western society, such stores could be in operation under government licensing within our lifetime. Your neighborhood parent store will probably be in a "medical park" next door to a gynecologist's office for convenience.

Does this sound something like Aldous Huxley's nightmarish *Brave New World?* You can find this book in almost every library now, though there was a time when it was banned in many communities.

Huxley intended the book as a satirical fantasy of impersonal science. I doubt he ever dreamed in 1932 that it would one day be so close to potential reality.

Huxley portrayed a world called Utopia, run by humanistic scientists and social planners and ruled by a Great Being called Ford. In Utopia traditional monogamous marriage, parenthood, and religious rites are practiced only by "Savages" permitted to live outside the new world state on a reservation. Citizens of Utopia conform to the ways of science and socialism from conception to death.

Brave New World opens with a group of students touring the Central London Hatchery and Conditioning Center. They see new life being fertilized, gestated, born, and behaviorally conditioned in assembly-line fashion. A single female egg produces from eight to ninety-six "buds." Each bud grows into a perfectly formed human embryo which is then placed into a great bottle containing the simulated environment of a mother's womb. While the unborn is gestating, future intelligence and caste are determined by the controlled flow of oxygen into the bottle. Those intended for lower intelligence and caste are given less oxygen.

Members of the lowest caste, called Epsilons, are born

virtually mindless. They wear black and are programmed to do menial labor. Gammas, dressed in green, are made a bit smarter to handle more complicated jobs. Khaki-clad Deltas are pushed another notch up the social ladder. Elite Betas form the upper crust. Alphas are the leaders of society.

Standardized people. Instruments of social stability. Strictly controlled population. As new humans emerge from the hatchery, senior citizens are painlessly eliminated by blissful whiffs of gas to keep the population stable.

During a citizen's lifetime, every physical need is satisfied. With procreation turned over to hatcheries, all sex is recreational. Everyone has everyone else sexually. Promiscuity is a sacred commandment.

The people of Utopia hold song fests in which they sing "Bottle of Mine" instead of "Mother of Mine." They take communion in the name of the Great Ford. Calendar years are numbered A.F.—"After Ford."

In 1946 Huxley wrote a new introduction to *Brave New World*. "It seems quite possible," he said, "that the horror may be upon us within a single century."

Huxley died in 1963. Were he living today he might agree with Dr. Paul Ramsey, Professor of Moral Theology and Ethics at Princeton Theological Seminary:

> There shall come a time when there will be none like us to come after us. . . . The Central London Hatchery will become a possibility within the next fifteen to fifty years. . . . Philosophers, theologians, and moralists, churches and synagogues, do not have the persuasive power to prevent the widespread social acceptance of morally objectionable technological "achievements" if they occur.[4]

In 1970 Ramsey was thinking of genetic engineering, which makes AID, IVF, and cloning seem like freshman biology. Gene engineers propose nothing less than rede-

16

signing the genetic blueprint of characteristics which predetermine the physical makeup of every person.

Genetic engineering holds undreamed of possibilities for curing over 2,500 known genetic diseases at their source. Hemophilia, cystic fibrosis, muscular dystrophy, Huntington's chorea, and other dread afflictions for which there is now absolutely no cure, could be wiped out. Some types of cancer and heart disease along with other broader ailments which are known to be genetically predispositioned are also included in the target range of the next twenty to thirty years. Some think that if industry and the environment can be cleaned up, eliminating cancer-causing agents, for example, a life-span of nine hundred years might not be unrealistic. A few promise that, barring accidents, *Homo sapiens* born in the next century will live forever.

Much of this is not as fanciful as you may think. Just as physical scientists are delving into subatomic particles that form the foundations of matter, genetic engineers are tinkering with the chemicals that "print" the blueprints of life. Already they've synthesized and recombined the basic life chemical, known as DNA (deoxyribonucleic acid), in some simple life forms. Experiments are now going on to implant good DNA in human cells to counteract bad DNA. The research is now in the elementary stage, but the "show is on the road" toward altering undesirable hereditary characteristics. How far will this go? Some say science will eventually be able to design man as he can only be idealized today.

The scientific community is divided over whether heredity-changing research should continue. More than a few researchers lie awake nights worrying that a fast multiplying killer microbe, developed in the lab from recombined DNA, might escape and infect multitudes of people. In 1973 scientists became so alarmed over this possibility that they actually called a voluntary halt to DNA experiments. The moratorium was not lifted until 1975 when safety

17

guidelines were drawn up and accepted. Many scientists still don't think the safeguards are foolproof.

Another concern is that science will only succeed in building a monster. Worriers point to mistakes of the past such as the frightening thalidomide disaster. The drug was originally sold outside the United States as a tranquilizer and taken primarily by women to help cope with discomfort during the early weeks of pregnancy. Thalidomide apparently worked havoc with the development of the unborn, resulting in hundreds of children born with deformed limbs—some with arms, for example, that resemble a seal's flippers.

There is also fear of the direction which the biorevolution is taking. Almost all of the leading research scientists in this field are admitted atheists or agnostics. They look only to man for ethical guidance, not realizing that "the fear of the Lord is the beginning of wisdom" (Psa. 111:10). They reject the God of the Bible. For them, man is alone in the universe, a product of evolution by chance, with no hope except in himself.

They hope for a disease-free super race. This is man-made evolution at the ultimate—ideal man as visualized by humanistic social planners. This man will see religion only as a vestige of his primitive past (an approach now taken in secular universities and some denominational church schools). A humanistic utopia beyond *Brave New World*.

Is this where humanity is heading? Is the biorevolution setting the stage for the biblical Anti-Christ to arise and proclaim himself God?

We don't know the answers to all these questions—yet.

For the present, the key question is: How should we respond as Christians to the bioethical revolution with its enormous potential for both good and evil? God may be saying to us as he said through Moses to the nation of Israel:

I call heaven and earth to witness against you today, that I have set before you life and death, the blessing and the curse. So choose life in order that you may live, you and your descendants (Deut. 30:19).

IT'S CLOSER THAN YOU THINK

After the smashing success of *Roots,* the fourteen-year-old daughter of a friend came home and announced that author Alex Haley had committed suicide.

"Why would he do that?" asked her gullible father.

"He found out he was adopted," the girl replied in mock seriousness.

Alex Haley is alive and well and collecting royalties from the book he did on his family line. And reference librarians are still up to their necks with genealogy requests from ancestor hunters inspired by Haley's dramatic story.

The great interest in genealogy triggered by *Roots* shows that nothing has yet come along to replace the sense of identity and belonging that a family ancestry provides. Cloning, as I'll explain a little later, will drastically alter the family lines and units which man has known since creation.

The concept of family, of course, is much more than names and accomplishments of long-dead progenitors. The discovery of ancestors is not always pleasant. One fellow I know discovered that his great-great-grandfather was a murderer. He also found a horse thief hanging from his family tree.

What I find more interesting is trying to figure out who got what from whom. For example, I have a peculiar hair pattern that makes it impossible for me to grow long sideburns. This same pattern has been observed in my family for four generations.

Patterns of eye color are also fun. Our sixteen-year-old Lee Ann has green eyes which were inherited through me, my mother, and my mother's father. But our four-year-old Jennifer has blue eyes. This blue resulted from the combination of genes in her mother's family and my family.

Every physical and mental trait you and I possess has been affected by the gene pool of our family heredities. Certain traits are more obvious among the immediate generations. But occasionally some oddity shows up that we can't blame on any ancestor we know. That's usually an inheritance from a more distant predecessor or is produced by something other than heredity.

Your family tree defies the cataloging of every branch. In ten generations you can count 1,056 grandparents, minus a few if any of your cousins married each other. It wouldn't take many more generations to reach a million grandparents. Imagine—your eye color, nose shape, skin texture, and countless other characteristics are determined by genes passed through thousands of grandparents, clear back to Adam and Eve.

This potpourri of traits makes up your unique genotype (hereditary blueprint). Unless you are an identical twin, there is nobody else in the world quite like you.

Identical twins are a type of clone because they share an identical heredity. They are produced by the division of the cell mass that grows from a single fertilized egg. Identical twins are fairly rare, occurring once in approximately 344 births in the U.S. We can learn a lot about the members of a clone by studying such twins, especially twins that were adopted as infants by different families. Take twins James A. Spring and James E. Lewis.

They were born in Piqua, Ohio, on August 19, 1939, and adopted a few weeks later by families living only forty miles apart. Neither family was told that their adopted son had a twin brother and the twins did not meet until almost forty years later.

They discovered some amazing coincidences. Both had been named Jim. Both had pet dogs named "Toy." Both had taken law enforcement training. Both pursued hobbies of blueprinting, drafting, and carpentry. Both had married first wives named Linda and second wives named Betty. Both had named their first sons James Allen. They were also physically alike, six feet tall, 180 pounds, had dark hair and brown eyes, and identical faces, as well as the same brain wave and heartbeat patterns. The results of all the psychological tests they took looked as if one person had taken the same tests twice.

Some identical twins have been known to experience the same illness and even die near the same time. Drs. Ian C. Wilson and John C. Reece tell of twin sisters who displayed the same symptoms of mental illness and were put in a room together at a mental hospital. When their condition worsened, attending physicians moved them to separate rooms. The stronger sister was found curled in a fetal position, dead the first night of separation. A few minutes later her twin was also found dead and lying in the same position. One of the psychiatrists described them as similar to "two people with one brain."[1]

Since I am a geneticist and not a psychiatrist or a parapsychologist, I can't explain this. I can only suggest that this offers us some preview of relationships between clonal people.

Identical twins have the same genetic "serial number" stamped in every cell in their bodies, as will members of a clone. This can be life-saving when one needs a transplant.

Twins Chris and Craig Bowman, whose parents were our fellow church members in Orlando, Florida, had such an experience. Ten years ago Chris began experiencing chronic backaches. His doctors removed a growth from his back, but he did not recover well from surgery. A more thorough investigation led to the diagnosis of lymphosarcoma, cancer of the white blood cells. Chris' one medical hope was a bone marrow transplant to arrest the growth of the cancer.

The big problem with transplants is rejection. The human body has intricate mechanisms that enable it to recognize foreign matter, whether disease bacteria or a donated heart. The body's defenses attack the foreigner as an intruder. Doctors counterattack with drugs to suppress the natural defenses, but this makes the patient more vulnerable to infectious diseases. The battle is touch and go and often results in rejection. Then the surgeons must look for a new donor.

Craig came to the rescue and donated some of his bone marrow. It was the perfect genetic match. Chris' antibodies regarded the new marrow as their own and did not fight the transplant. The cancer was stopped and Chris was healed. Today he is in good health and happy as the new assistant pastor at the Baptist church in Latta, South Carolina.

Chris had only one identical twin to provide help. A member of a clone may have several such "twins." In time of need any one of them could provide a transplant of identical tissue. Because of this, the possibility exists that a powerful person might have himself cloned just to have a supply of young donors available when he needs new body tissue or a new organ.

Twins are most similar to an individual and his clonal offspring. They differ only in their relative ages. Both twins and clonal offspring are the product of fertilization; you might say that what really differs is when they split—in the case of twins after a matter of hours, in the case of cloning after a matter of years.

A case could be made that, just as twins share the same parents, so the cell donor and his clonal child have the same parents. The cell donor actually is no more the biological parent of his clone than is one identical twin the parent of the other. To put it another way, the cell donor should experience emotions of fraternity, rather than those of paternity toward his clonal offspring.

To understand the strange phenomenon of cloning we need to take a brief look at basic genetics.

We once thought that the cell, the basic unit of life, was a simple bag of protoplasm. Then we learned that each cell in any life form is a teeming microuniverse of compartments, structures, and chemical agents—and humans have billions of cells.

We call the hereditary train-like specks, which swim in the cell nucleus, chromosomes—meaning "colored bodies," from the way they absorb laboratory dye. The "passengers" which they carry are genes and genes are made of the vital chemical DNA.

Every living thing has a specified number of chromosomes in every body cell. A corn cell has twenty, a mouse forty, a gibbon forty-four, and a human cell forty-six.

The forty-six chromosomes in a normal human body cell come in two matching sets of twenty-three each, with each parent providing one set. The chromosomes are numbered one to twenty-three. There are two number twelves, for example, one from each of the parents. To have more than two of a number is abnormal and almost always results in tragedy.

In fertilization the father's sperm cell with twenty-three chromosomes unites with the mother's egg cell containing the same number. This produces a single cell with forty-six chromosomes which starts dividing. In a few days there is a discernible embryo. The hereditary blueprint, product of the parent heredities, was there from the instant of fertilization.

Unfortunately, every new blueprint is flawed by genetic "mistakes" accumulated and passed down the family lines. Some mistakes produce one or more of the 2,500 known genetic diseases, which include hemophilia (in which a critical blood clotting substance is absent, leaving the victim in danger of bleeding to death from a slight injury or bruise), Down's syndrome (mongolism) and a predisposition to certain forms of heart disease and cancer. Some result in less troublesome problems such as flat feet and color blindness. No person is completely free of genetic mistakes. It is merely a matter of degree and combinations. Everyone carries at least a few problem genes in his chromosomes.

Genetic mistakes are called mutations, alterations of the genetic blueprint. This may involve the unnecessary duplication of a chromosome, as in Down's syndrome, in which there are three number twenty-ones instead of two; detachment of a portion of a chromosome which then hooks onto an unrelated chromosome (called translocation); "typographical errors" in copying the genes for the next generation and other errors.

Every human has developed from the genetic blueprint bequeathed by the combined heredities of his parents— except for Jesus Christ, the virgin-born Son of Mary whose Father was God. Scripture says this plainly. It was predicted by at least one Old Testament prophet and accepted as fact by the New Testament writers.[2]

Because cloning is a sort of virgin birth, we need to consider the unique nativity that occurred in Bethlehem.

Unbelievers, of course, say the biblical writers papered over the real story. One theory is that Mary was impregnated by a soldier from the Roman occupation army. Another says that Joseph and Mary let passion overrule while they were engaged. Both theories are illogical. For one thing, it seems incredible that Jesus' enemies would not have learned about such a major violation of the Law and used it to refute Jesus' claim of deity.

Still another proposal is that the "virgin birth" was a freak biological happening, similar to the parthenogenesis

(Greek for virgin birth) that occurs in much of the plant and animal world.

By the laws of genetics this is impossible for humans. Animal virgin births involve doubling the chromosomes in a female egg to give the egg a full number of chromosomes and the ability to develop into an adult. Mice, for example, can be bred in a laboratory to become genetically pure, that is, without harmful mutations. Doubling their chromosomes would be fine. Humans, however, breed by choice and are far from genetically pure. Through generation after generation, the human gene pool has accumulated a vast amount of genetic garbage. To double the chromosomes in a human egg would result in a double dose of harmful mutations and almost certainly be lethal. Doubling through normal fertilization by a sperm usually does not result in mutations "stacked" one on another. The differences in the two heredities prevent this.

But let's suppose that Jesus was born this way and that he survived "almost certain" genetic death. He would have been a genetic cripple, retarded, and grossly deformed.

He would also have been a female.

Follow closely, and I think you'll understand. One of the chromosomes in the human egg is always an "X" in genetic language. The sperm cell may have either an X or a Y chromosome. If the sperm that fertilizes the egg carries an X, the resulting zygote (fertilized egg) will have two Xs and be a female. If the sperm carries a Y, the zygote will be male. If Mary's chromosomes had doubled to produce Jesus, "he" would have had two X chromosomes and would have been a "she."

Then how could Jesus have been born without a human father? God could have created a special Y sperm, with twenty-three perfect chromosomes, to fertilize one of Mary's eggs. Since I accept God as the Creator of the universe and of each "kind" of life, I have no difficulty believing that he used this method or some other means beyond his usual law of reproduction to accomplish this biological miracle. " 'For My thoughts are not your thoughts, neither are your ways My ways,' declares the Lord" (Isa. 55:8).

Cloning was definitely not the way Jesus was conceived, although the non-human life forms that achieve a virgin birth are, in a sense, clonal offsprings. These include honey bees which reproduce both bisexually and asexually. But only the male drones are produced from unfertilized eggs. Female workers and the indispensable queen bees are born from fertilized eggs.

Science has long known how to make simple life forms reproduce asexually. My first experience in this was as an undergraduate at the University of Florida. In a microbiology lab we produced clones of bacteria easily. We first isolated a single bacterial cell. Then we made that cell grow and divide in a culture tube to produce millions of cells. These millions of cells, all copies of one another, made up a clone.

In recent years cloning has been accomplished with higher life forms that do not normally reproduce asexually. Frogs have been cloned, mice are in process, and research on human cloning is going forward. The cloning of frogs and mice involves somewhat different techniques.

The first step in cloning frogs is to obtain a female ripe with egg cells. The eggs are removed with a skillful squeeze, then exposed to a strong light. This causes the eggs to lose their nuclei. Remember the nucleus of a cell, including an egg cell, contains the chromosomes, the carriers of heredity. An egg without a nucleus is therefore a cell with virtually no heredity, with no blueprint or instructions on how to develop into a frog.

The scientist next removes a tadpole from his aquarium and cuts out the intestine. He then extracts a nucleus from an intestinal cell which he inserts into a nucleus-less egg.

He can repeat this process with other nuclei for the number of frogs he wishes for the clone.

The eggs with their new nuclei develop into tadpoles—all identical twins to that poor tadpole who gave his life for science. The tadpoles progress into adult frogs, each a "carbon copy" of the others.

Cloning mice is quite different. The scientist treats the mouse egg in such a way as to double the number of chromosomes in the egg nucleus. Each mouse cell, as I previously mentioned, has forty chromosomes with one exception. Female mouse eggs and sperm, in the case of a male, have only twenty. At fertilization, a twenty-chromosome sperm cell fuses with its egg counterpart to get the full complement of forty, just as happens in human reproduction when the twenty-three chromosomal quota is doubled.

The trick with mice is to double the chromosomes in a mouse egg cell without fertilization by a male sperm. If this can be done and the egg cell with forty chromosomes placed back into the female uterus, then presumably clonal reproduction will result and the offspring will be genetically identical to the mother. Because mice have short reproductive cycles, harmful mutations can be eliminated by controlled breeding. The "stacking" of chromosomes will not create problems.

I've already noted that the stacking of chromosomes will not work in humans. If this is still unclear, please be patient. It will be explained more fully in later chapters.

The method used in frog cloning appears to hold the most promise for human cloning. Researchers persuade women patients having abdominal surgery to donate eggs. They extract the nuclei from the eggs, leaving the rest of each egg intact but robbed of virtually all its heredity. But instead of taking intestinal cells, as they do from frogs, the scientists collect special cells from male patient volunteers having surgery on their sex organs.

These cells are "parent" cells for sperm cells. In the male they are extracted from the testes. Each bears a full order of forty-six chromosomes. It is the sperm cells and egg cells that contain only twenty-three.

Dr. Landrum Shettles, one of the foremost biological scientists living today, has reportedly made substantial progress with sex cells. Three times, he reports having successfully removed the nucleus from a female egg cell and inserted a parent cell. Each time the egg with its implanted nucleus of forty-six chromosomes grew and divided until it reached the multicelled stage when, according to Dr. Shettles, it was ready to be implanted into a woman's uterus using the process perfected for test-tube babies. Dr. Shettles claims he halted the experiment at this point. I presume this means he killed and preserved the human cell masses for future study. He said, "There was every indication that each specimen was developing normally and could readily have been transferred into a uterus . . . where it would have developed into an identical twin of the man who donated one of his sex cells."[3] If Dr. Shettles has accomplished what he claims, the first human cloning may be near at hand.

Of course, there are skeptics in the scientific community. They note the extreme difficulty of getting a microscopic cell into an enucleated egg cell. A human egg cell is only 1/250 of an inch in diameter, much smaller than a frog cell.

Even if it could be done, the doubters say, there might be a destructive interaction between the transplanted nucleus and the cell cytoplasm in its new home. The cell could refuse to grow and divide or it could develop abnormally.

A third major obstacle involves the genetic destiny of body cells. Some cells are destined to build fingernails, some to construct nasal passages, and so on. Yet each cell continues to hold forty-six chromosomes with blueprints for the whole body. But the sex cells which Shettles claims to have used in his experiments are not limited in their destiny, because their offspring, sperm cells and egg cells, must be capable of producing all kinds of cells. The use of body cells which have already been "switched on" for special jobs is far more unlikely, in my judgment, to be successful in human cloning.

What are some other opinions? Dr. Kurt Hirschhorn, chief of the division of medical genetics at Mount Sinai School of Medicine in New York City, is a former president of the American Society of Human Geneticists. He thinks human cloning is not only possible, but probable. Dr. Bentley Glass, Distinguished Professor of Biology at the State University of New York, projects "C-Day" by the end of this century. Dr. James F. Bonner of the faculty of the California Institute of Technology said as far back as 1968 that within fifteen years science will know how "to order up carbon copies of people."[4] Science writer Isaac Asimov says, "Biologists are sure that some day they will be able to clone mammals, and even human beings."[5]

It's interesting that David Rorvik, author of *In His Image,* and Dr. Shettles, the scientist who claims to have developed a clonal embryo, are good friends. In 1970 they coauthored a book on how to choose your baby's sex before birth. Rorvik has used Shettles as a source for many of his articles.

Rorvik claims in his account that the first human was cloned in December, 1976. He shrouds the experiments in mystery by giving make-believe names to his characters and by being less than totally specific about the methods used. However, I must concede that his scientific facts and reports are right, indicating that he has researched the subject thoroughly.

The friendship of Shettles and Rorvik and the fact that Shettles claims to have partly succeeded induces the expectation that Shettles might have been the mysterious "Dr. Darwin" of Rorvik's book, who cloned the multimillionaire "Max." Rorvik states positively that Shettles was not the scientist. He also calls the book a mixture of truth and falsehood.

The name "Darwin" is a key to the philosophy of evolutionary humanism that is behind the book and the research into cloning as well. Rorvik decides moral questions by consequentialist ethics, another way of saying the end justifies the means. Using this reasoning, Rorvik justifies "Dr. Darwin's" deception of the women used in experiments, which involved exchanging nuclei from eggs with nuclei from body cells, tricking their reproductive processes, and forming clonal cells.

Even if Rorvik's story is a fabrication, I still predict human cloning is coming soon. Many different lines of research are converging to give us the capability of making genetic copies of ourselves. Just as Hitler's doctors, using consequentialist ethics, were able to justify heinous experiments for useful information, Rorvik has shown that it is possible to justify cloning if the right people want it to happen.

Human cloning will not be attained without much trial and error and the destruction of numerous egg cells and human embryos. Some of the embryos may develop into cloned "monsters" before they die or are killed in the laboratory. Steptoe and Edwards admit many failures before producing the first test-tube baby. Others who have worked on IVF embryo transplants have dumped the remains of their experiments into the laboratory garbage for burning in an incinerator.

Rorvik admits that his "worst doubts about cloning concerned the issue of 'bench embryos,' . . . experimentally conceived in the laboratory and then sacrificed in the process of carrying out some research procedure."[6] Apparently Rorvik overcame his qualms by use of consequentialist ethics.

I'm pessimistic about the possibility of any great public outcry when the first human cloning takes place. We hear much praise for the Steptoe and Ewards test-tube baby.

Who speaks for the embryonic lives that were destroyed to perfect the technique which allowed Baby Louise Brown to be born?

A million unborn babies are killed in the U.S. each year. Every public hospital with a maternity ward has the equipment to dispose of human embryos and fetuses. The administrators and doctors and nurses may abhor having to do abortions, but they have no choice, short of disobeying 'he Supreme Court ruling of 1973.

What difference will a few hundred or thousand clonal embryos and fetuses make to humanistic scientists and government bureaucrats?

There are others beside right-to-lifers, fundamentalist Protestants, obedient Catholics, and Orthodox Jews who are against cloning. Embryologist Robert T. Francoeur, author of *Utopian Motherhood,* is one of several world-renowned scientists who are objecting. "Xeroxing of people?" he asks. "It shouldn't be done in the labs, even once, with humans."[7]

These voices are drowned out by other scientists who seem to believe that researchers should have a carte blanche in experimentation. One of the most vocal is Stanford University's Dr. Joshua Lederberg, a Nobel Prize winner. He declares, "If a superior individual—and presumably, then, genotype—is identified, why not copy it directly. . . ."[8]

Whatever the future holds, we cannot hold our peace and do nothing. "The potential for human cloning," says Jeremy Rifkin (coauthor of the book *Who Should Play God?),* "no matter how far in the future, challenges our entire value system. We must talk about the implications now, before any crisis occurs."[9]

I wholeheartedly agree. The issue of cloning, however, has additional implications for the Christian. We will consider these in the next chapter.

3

WILL
CLONES BE
SUBHUMAN?

A handsome young man we'll call Bill asks to join your church. Bill has no genetic mother. Because there was no mixing of genes from two heredities at his conception, Bill is his father's twin. He is truly a "chip off the old block." He not only looks like his father, in a hereditary sense he is his father with no chance of substantial variation.[1] Bill is a clonal man.

Should Bill be cloned, his offspring will be both in his image and in his father's likeness. This will continue through all future clonal generations. The ancestry of Bill's clonal descendants will not be a family tree, with branches running off the trunk, but a straight pole. If Bill's father and twin has blue eyes and a hooked nose, all of the clonal posterity will have blue eyes and hooked noses.

Cloning will work the same way laterally. All of Bill's brothers will be identical twins.

With Bill's request for membership, your church faces the question: Does a clonal human bear the spiritual "image of God"? Does he have the capacity to know and worship God? Will he, like ordinary humans, be a sinner by nature and by choice? Or will he be amoral, neutral, and not responsible for his own acts?

These questions do not concern those who see man as only a happening of cosmic chance which evolved from a bubble of amino acids into mechanistic life. They do challenge those who perceive man as more than chemistry and biology. Man is soul and spirit, we believe, made in the image of a loving, all-powerful Deity. Though marred by a sin nature acquired at the Fall and expressed in rebellion, self-seeking, and self-exaltation, he is sought by God through the atoning, redemptive sacrifice of Christ.

We must ask now: Will Bill and other clonal children be subhuman, lower than ordinary man, but higher than apes? Will God love clonal *Homo sapiens* as much as he loves those with a genetic inheritance from two parents?

We must be thinking now where clones will stand in the order of God's creatures. We cannot afford to be caught unprepared when cloning does happen. We should begin now to shape the ideas and the ethics of the future and not leave the field to unbelievers. The spiritual dignity and value of man in western society may be at stake.

Our basic sourcebook is the Bible. We begin in the Genesis record, which is neither fable nor "true myth" but authoritative and true.

The focus in chapter one is on the entire creation process. God is Planner, Architect, and Contractor. Earth and seas, herbs and trees, fish and fowl, animals and finally man, the crowning touch, are created at his command.

He created all life forms by "kinds" (Gen. 1:21, 24, 25). From these kinds came the species which we know today. The cat "kind," for example, includes everything from tabbies to tigers, and the horse kind ranges from Shetland ponies to zebras. The original members of each kind were evidently given a great deal of genetic variety. As each kind obeyed God's command to "be fruitful and multiply," the processes of reproduction produced within each kind a number of species.

Creationists hang creatures by kinds on separate an-
cestral "trees." Evolutionists drape all living forms on
one big tree and claim that all evolved from the accidental
explosion of life from dead matter at the tap root. (Inci-
dentally, I was a Christian evolutionist—believing God
was behind the evolving—until after receiving my doc-
torate in genetics. Upon being challenged by Dr. Duane
Gish of the Institute for Creation Research, I took a year
to sort out the evidence. I concluded there is more scientific reason to believe in special creation than in evolution.)

The second chapter of Genesis focuses on man and the
environment which God prepared for him. The key statement is Genesis 2:7: "Then the Lord God formed man of
dust from the ground, and breathed into his nostrils the
breath of life; and man became a living being." This can
be interpreted two ways. One is that God breathed ani-
mate physical life into man. The Hebrew word for "being" (translated "soul" in the King James Version) is
nephesh. It is used in some other Old Testament passages
for the life principle essential to the existence of animals
and man. The other rendering is that God breathed into
man an eternal spirit which animals do not possess.

I prefer the second solution. God is Spirit and the life he
breathed into Adam was spiritual life.

Scripture speaks of the spirit of man hundreds of times.
The word has different meanings in respect to contexts,
but in general the connotation is that the spirit of man is
distinctive and unique among God's creatures. Spirit is the
moral and spiritual essence that is in God's image and is
not limited by the physical form which returns to the dust
of the earth at death.

Spirit is the channel through which God communes with
man. So Jesus told the Samaritan woman, "God is spirit;
and those who worship Him must worship in spirit and
truth" (John 4:24). No animal has the spiritual capacity to
know and worship God.

37

Will a clonal offspring, formed in the genetic image of a single human parent, be such a spiritual being? The creation of woman may give us some hint.

> So the Lord God caused a deep sleep to fall upon the man, and he slept; then He took one of his ribs, and closed up the flesh at that place. And the Lord God fashioned into a woman the rib which He had taken from the man, and brought her to the man (Gen. 2:21, 22).

This was the first use of anesthesia and the first surgery, with God acting as both anesthetist and surgeon. Was it also the first cloning, since Eve was made from a single parent?

By the laws of genetics as we know them today, she could not have received her total genotype from Adam, or else she would have been a male. Clonal reproduction always produces the same sex.

We've said that a male is XY and a female XX in genetic parlance. Using Adam's rib as the raw material, God may have eliminated the Y chromosomes and duplicated the X chromosomes in each cell. This would be the minimum genetic reconstruction required. More likely, God completely reorganized the molecules of life into a brand new genotype for Eve. In any case, Eve was created before the reproductive cycle by sexual union began, with its mixing of parental genotypes to produce new genetic identities.

This leads to a thornier question: Did Eve receive her *imago dei* through Adam or was she given her spiritual identity during the "surgical" procedure?

Not being all-wise, I can't answer that. If the former is true, it could indicate that each person is "stamped" with God's image through or from his parentage when he comes into physical existence. A clonal person then would bear the image of God as much as anyone else.

Will a member of a clone be responsible to God for his

moral failings? Is he included in God's plan of redemption?

Again, let's look at the first couple.

Adam and Eve were created perfect. Not one of the millions of genes in the forty-six chromosomes of each of their billions of cells was blemished by a harmful mutation. Their environment was also untainted by the pollution which plagues us today. Their bodies would have suffered a lot of wear and tear, but in their primeval perfection, the life processes might have followed a renewal cycle that would have enabled them to live forever.

However, they listened to the tempter and disobeyed God by eating of the forbidden fruit. God had said, "You shall not eat from it or touch it, lest you die" (Gen. 3:3). This meant spiritual death and physical death as well if God had originally intended that their bodies last forever.

Immediate results of the first sin are presented in the Genesis account of the Fall. Both Adam and Eve felt guilty and tried to hide from God. They were cast out of Eden. The serpent was told there would be "enmity" between his seed and Eve's seed. "He [the seed of the woman] shall bruise you on the head," God informed the serpent, "and you shall bruise him on the heel" (Gen. 3:15). Many theologians call this the first Messianic promise, a foreshadowing of the conflict between Jesus and Satan, culminating in Jesus' ultimate victory over the devil.

Eve will bear children in "sorrow" (the pain of childbirth), the Genesis account records. The man will "rule" over her (and women have been exploited by men ever since). The ground under them is "cursed." Adam must sweat to make a living among the thorns and thistles until he returns to the ground in death. This indicates that if Adam and Eve had not sinned their physical bodies might have endured forever.

Sin spread through Adam and Eve's posterity. "Therefore, just as through one man sin entered the world, and death through sin, and so death spread to all men, because all sinned" (Rom. 5:12). Scripture characterizes sin as the universal malady of the race, involving transgression, going astray, missing the mark set by God, and willful rebellion. "All of us like sheep have gone astray, each of us has turned to his own way" (Isa. 53:6). "There is none righteous, not even one . . . for all have sinned and fall short of the glory of God" (Rom. 3:10, 23). The only remedy is the atoning, substitutionary death of Christ. God "made Him who knew no sin to be sin on our behalf, that we might become the righteousness of God in Him" (2 Cor. 5:21).

There is no reason to believe that clonal people will not be sinners as we are. I don't think there are any genes for sin, yet it is a fact that sin flows from one generation to another. Learning to sin from association with others is not sufficient explanation. Kids grow up doing what comes naturally—sin. What is more natural than sin? Members of a clone will sin like everyone else. They will be in need of redemption. This should be reason enough to include them in God's circle of redemptive love and accept those who believe into our congregations.

Back to the matter of spiritual identity. When will a clonal life become a person with unmeasured value in God's sight? Reincarnation (identity in past lives) is all the rage in the circles of pseudo-sophisticates today. At the moment actor Steve McQueen is reported to be telling friends that he was Moses in a previous life. The Bible doesn't teach reincarnation. This is just another deception of the devil, enabling some to believe that they are part of some eternal continuum. Neither the identity of clonal persons nor that of people of ordinary birth can be submerged in the generations, past or future. Every individual has a self-identity. Genetic likeness will not merge two into one.

How does spiritual identity arrive? What or who is the conveyor? Is it borne along through the reproductive channels of the generations? Is it related to heredity? Are two parents necessary?

These are sufficient questions to keep Christian theologians and philosophers debating for a thousand years. I don't have any final answers. I can only say that one is always a complete being—body, spirit, and soul. He is never a half person.

Exactly when during the conception-implantation-gestation-birth process does one become a soul? This is relevant to cloning, test-tube babies, and other new methods of reproduction and childbirth. It is critical in the abortion controversy.

The Catholic Church officially holds that the soul or spirit is infused at the moment of conception. Original sin, Vatican authorities dogmatize, is also present from this instant. Thus Catholic dogma outlaws IUD "birth control" devices and morning-after pills which destroy a fertilized egg.

As you might expect, Protestant theologians and doctors disagree among themselves. The most conservative agree with theologian Francis Schaeffer and pediatric surgeon C. Everett Koop who say that personhood is attained at conception. Dr. Thomas Monroe, an obstetrician-gynecologist and Presbyterian lay leader in Chattanooga, and others take a more modified stance. They believe the soul is given at the time when the fertilized egg has "slid" down the mother's fallopian tube, developed into an embryo (blastocyst), and implanted itself in the wall of the uterus. The position of pro-abortion liberals is that the new life is not a full person until born with the capacity to sustain itself. I wonder what the liberals think about "preemies" who cannot survive independently, but can be sustained by modern medical technology.

Personally I think the implantation argument is an escape mechanism for early abortion. Biological life truly begins at conception since all the genetic blueprints for the baby's future development are in the fertilized egg. I can't

41

see how implantation makes a substantial difference, except that the new life has a much greater chance of succeeding. Is it really any different from the life in the fertilized egg? I think not.

However, we run into some real problems with cloning if we say that a person exists spiritually as well as physically at the moment of conception. Identical twins, which are members of a clone, materialize not at conception, but a number of cell divisions later, when the mass of cells breaks apart to form two identical and separate cell systems. If the spirit was given at conception, then the twins might have to share a common spirit or one might do without a spirit altogether. A clonal offspring, you remember, is also an identical twin to his parent.

Furthermore, there is at least one documented case in which two fertilized eggs developed and fused to produce one person. The woman is now living in Vienna, Austria. Scientists at the University of Vienna, who discovered the freak birth, call the woman a genetic mosaic. Her parents had produced two fertilized eggs, as happens with fraternal twins, but early in development the embryos merged to form one fetus. Cells from one of the fertilized eggs made the woman's blood and skin cells, while cells from the other eventually produced her eggs.[2]

I expect this may have happened elsewhere, but the phenomenon went undetected. When two fertilized eggs become one, is there a doubling of the spiritual dimension in one person?

We're getting on the edge of theological absurdities. I'm satisfied to leave these unexplainables to the wisdom of God. If we knew everything, we would be God. We just don't know how or when a human becomes a spiritual being.

God may not work by our rules at all. It's that way in physics. Dr. John H. Martin, a researcher at Argonne National Laboratory near Chicago, observes that an electron in the nucleus of the atom can be made to go through two different places at once. "This is impossible in our familiar world," he notes. "I know that my boy playing baseball cannot bat a fly into both right and left fields at the same time." Yet an electron can go two places at once! What does a physicist do when he meets the "unreasonable"? Dr. Martin says, "We always assume . . . that it is our understanding that is at fault. . . ."[3] Biologists should be just as reverent.

Though we can't pinpoint the moment an embryonic human receives a new spiritual identity, this does not mean that man is not a spiritual being. There are many evidences that man is vastly different from apes, orangutans, dolphins, and other "smart" animals.

If we can't be so definitive about humans conceived and born according to the laws of nature, I doubt that we'll do any better with cloned humans. Given our deficiency of wisdom, I think we should be prepared to award full personhood to all cloned beings.

We won't be able to tell members of clones from those born the ordinary way. If we are to believe David Rorvik, there's already one clonal offspring walking around. From what we know about animal cloning, a cloned human will have all the human characteristics of speech, thought, senses, and physical appearance. He'll be as different from an ape, for example, as you and I are.

If I were on the church membership committee, I would vote for Bill to be accepted into the local Christian fellowship if he professed faith in Christ. I would accept him on the same basis as I would anyone else. The fact that he is both the son and twin of his father should have nothing to do with his relationship to me and to our Christian brothers and sisters. Bill will be as human as my own son.

Yet, I still believe that humans should not be cloned. There are very good reasons for this, both from a scientific and a Christian perspective. We'll consider these in the next two chapters.

SHOULD
HUMANS BE
CLONED?

Scientific meetings aren't always the stuffy sessions they're made out to be. At a March 1977 forum of the National Academy of Sciences in Washington, the subject was cloning. A learned researcher was lecturing when a band of objectors suddenly began chanting, "We shall not be cloned!" Swarming onto the stage, they unfurled a banner proclaiming: *"We will create the perfect race— Adolf Hitler."*

Scientific debate about cloning is usually a little more orderly, but still spirited. Rarely will you find anyone neutral about whether a human should or should not be cloned.

Molecular biologist Leon Kass is executive secretary of the Committee on Life Sciences and Social Policy for the National Academy of Sciences. He told an annual meeting of the Association of American Law Schools in Chicago that a man should not be cloned even once. The ability to do so, he said, is not a justifiable reason. Princeton's Dr. Paul Ramsey, the theologian often quoted by those against cloning, says prospective mishaps in cloning experiments, resulting in the formation and destruction of malformed embryos, calls for a moral prohibition.

Stanford's Dr. Lederberg, whom I've quoted before, and Joseph Fletcher, the liberal theologian, appear to be the main cheerleaders for cloning. Lederberg doesn't advocate doing away with the old-fashioned way of having babies, but he does think clonal reproduction should have equal opportunity. Fletcher says man himself must be in control of human evolution.

This is typical of the way evolutionary humanists think. Humanity, they believe, is not here by design and special creation but only by evolutionary chance. Since there is no Creator, Sustainer, and Controller at the head of the universe, man must look out for himself and work out his own destiny.

In this chapter I'm going to give the humanists a turn at bat. Then after each "ball" they hit for cloning, I'll have a turn to present my scientific point of view.

1. *Cloning is a great way to perpetuate genius.*

How about a hundred Einsteins, two dozen Picassos, and ten or twelve Tillichs? I don't mean that these gentlemen will be exhumed from the dead and their bodies searched for live cells. The way we bury people, the cells expire pretty fast. The ancient Egyptians did a better job. I've heard that it might be possible to reconstruct King Tut by cloning. More seriously, the promoters of cloning advocate selecting a few of the greatest now-living *Homo sapiens* for cloning.

Even if an exact chip off the old block can be produced, however, there is no guarantee that Einstein, Jr., will be as smart as his father. Scientists continue to debate the relative powers of nature and nurture in shaping a life. Is heredity or environment more important? Natural intelligence or the motivation to learn? We all know gifted youth who choose to dissipate their endowments through laziness, worthless hobbies, drug use, or some other desecration of their talents. Furthermore, the clonal son or daughter of a genius would live in another era. The milieu during which one lives affects the way he responds to challenges set before him.

Numerous studies have shown that accomplishments run in families. Children of doctors are more likely to become physicians, for example. Children born and nurtured by parents who value learning are more apt to graduate from college. There are exceptions, of course, but in general this is the way it is. It's an ego blow to educators, but some studies have shown that the home has more influence on a child's development than the school.

Unless adopted, a clonal child wouldn't have a normal family life. How would he fare in peer relationships? The parent of the clone might decide to keep his little group of alikes together. How would it be for three or four Einsteins to grow up in the same household?

Suppose the clonal child knew that he was the offspring of a genius. How would he be affected by the pressure to measure up? Any youth counselor can tell you about the agony which some kids face in trying to measure up to parental expectations.

There's another aspect of this proposed cloning of genius that concerns me. Who would select the candidates? If the cloning were to be done by a government or university team of biologists, a committee would probably be named. What criteria would they use? Would Christian faith and heritage be a plus or minus? Would they reject Francis Schaeffer and accept Joseph Fletcher?

Right now evolutionary humanists are pretty much in control of our higher educational system, the universities, the textbook publishing houses, and the government bureaucracy which funds scientific research. I doubt if they'd like a dozen more Francis Schaeffers or a hundred replicas of members of the Creation Research Society to which I belong. They certainly wouldn't want another Anita Bryant or Jerry Falwell.

You'll find a bias against fundamentalist Christians at

some graduate schools. I have a creationist friend who was afraid to publish his views on the origins of man until he got his Ph.D.

Dr. Robert L. Herrmann, Professor and Chairman of the Department of Biochemistry at Oral Roberts University Schools of Medicine and Dentistry, served on the admissions committee for the Boston University Medical School. He soon learned that two types of applicants were being rejected most frequently—fundamentalist Christians and Orthodox Jews. His colleagues seemed to feel that strong religious beliefs were disqualifiers.

There's a related issue to the cloning of genius which I haven't mentioned. This is the possibility of a megalomaniac having himself cloned. Or a mad scientist deciding to clone his beloved despot.

If you want a graphic picture of the horrors this could bring about, read Ira Levin's novel, *The Boys from Brazil,* published by Random House.

The book opens with a bizarre series of murders of adoptive fathers. Each victim is a civil service worker. As the plot unfolds, investigators track down a German woman, Frieda Maloney, who was hired by the organization to get a job with an adoption agency so she could look at their files. Following instructions, she selected applications of would-be adoptive parents indicating that the husband was born between 1908 and 1912 and the wife between 1931 and 1935. Each husband had to have a civil service job and both parents had to be white baptized Christians with a Nordic racial background.

At regular intervals the organization gave her babies to place with these families. She thought they were the illegitimate children of German girls and South American men and innocently went along with the scheme.

Actually they were the progeny of Hitler, cloned from cells taken from his body before his suicide and preserved for cloning by a mad Nazi physician, Dr. Mengele. The babies, ninety-four little Hitlers, were conceived in Dr. Mengele's lab and carried to term by Brazilian Indian tribeswomen.

Dr. Mengele planned for them to have parents similar to the Fuehrer's. Hitler's father was a civil servant, a customs officer. He was fifty-two when little Adolf was born; his wife, twenty-nine. The father died at sixty-five when the boy was almost fourteen. In the book, the adoptive fathers were being killed systematically about the time of their sixty-fifth birthdays when their adopted sons were almost fourteen.

In a chilling scene near the end of the novel Dr. Mengele meets one of the boys. "You were born from a cell of the greatest man who ever lived!" he shouts. "Reborn! . . . You're the duplicate of the greatest man in all history." I don't want to give the rest of the story away, but you'll find the boy's response intriguing.

The idea of cloning genius is not heaven-sent. It comes from another source.

2. *Cloning can provide soldier and servant classes of people.*

Joseph Fletcher suggests that top soldiers might be cloned to fight clonal soldiers of a tyrannical power. As far out as this sounds, I find it interesting that he admits that despots will use cloning. This isn't science fiction. If one person can be cloned, an army can be so produced. Presumably, the parents would all be fine physical specimens with keen minds and quick reflexes.

Soldiers are not the only special group which clonal crusaders have in mind. Special clones could be produced for other jobs which ordinary humans find distasteful or dangerous. Some universities are so hard up for football talent they might even guarantee alumni that funding for cloning projects would result in winning teams.

We would be cloning people to be used as slaves and work animals. They would look like ordinary humans, yet be treated as subhuman. They would have no more rights than animals. I can't imagine any humane person backing such a program.

3. *Cloning can improve the human race.*

Hitler's scientists probably never considered human cloning possible when they set out to purify the Aryan race. The way they chose was to breed the best male and female specimens available. If they had known how to clone, they certainly wouldn't have gone to the trouble of bringing in young Aryan men and women for sexual mating.

On the other hand, they might have preferred the old-fashioned way if they had known very much about genetics.

Cloning will bring about a genetic disadvantage for race improvement, depending on how much the procedure is used. The cloning of millions of people will diminish the variability resulting from the heterosexual mixing of genes. Generally, the combining of two different heredities will produce a stronger progeny. For this reason a farmer buys hybrid seeds instead of using his own seed from a single stock.

There is another problem which will become clearer later. Everyone possesses harmful genes. Cloning will keep these genes in circulation. Couldn't persons with problem genes be identified and classified "4-F" as draft rejects were in World War II? Probably not, unless some physical evidence of genetic disease was present. That type of person wouldn't be considered anyway.

Dr. Lederberg concedes that wholesale clonal reproduction will lead into an "evolutionary cul-de-sac" with no chance of genetic improvement. Ten years ago he proposed "tempered clonality," which will allow for both cloning and heterosexual reproduction.[1] He thinks selected cloning could improve the race by adding better people.

This will require a government regulatory agency with two major responsibilities: one, to select the candidates for cloning; two, to keep the clonal people from mating with members of other clones or the offspring of heterosexual reproduction. Allowing clonal people to mate would destroy any racial "benefits" to be gained by the selective use of cloning.

With human nature as it is, I don't see the latter as possible. A member of a clone will have the same urges as do all humans, and he will find a partner. His transgression would be found out when the child was born. Would the violators be punished for breaking the law of the clones? Would the innocent child be killed?

4. *Cloning can prevent genetic disease in a selected posterity.*

Again, this argument sounds good on the surface. The genetic engineers would simply pick out the best "stock" for cloning. Genetic disease caused by the mixing of certain heredities would be avoided.

This is not as fail-safe as it might seem. Members of clones will still be susceptible to mutations or mistakes in the replication of cells. They will also be affected by environmental influences which may bring out previously unknown genetic disease.

One of the many genetic diseases which can remain undetected until activated by environment and life style is Herniske-Korsakoff syndrome. An enzyme called transketolase, which regulates or filters vitamin B-1 to the brain, is missing from the cells. People with the defect don't get enough B-1 from their diets. The results are seen mainly in heavy drinkers of Northern European descent. At a certain level of alcohol absorption, the brain doesn't have enough B-1 to prevent the disorder from spreading. Without B-1 supplements, the victims go hopelessly insane.

Cloning will merely perpetuate such genetic defects. It

offers no panacea for the elimination of genetic disease.

5. *Clones can exchange body parts and experience enhanced social communion.*

This is one of Dr. Lederberg's key planks in his platform for cloning. Since all members of a clone are identical twins, a number of donors could be available for organ transplants. As identical twins, they can share a closeness and understanding that will facilitate harmony and peace. They will be able to work together cohesively.

This assumes that members of a clone will be ready to make the sacrifices required in donating organs. It assumes they will be free from jealousy, greed, and other failings of ordinary men. Utopian experiments have always failed to create perfection among ordinary people. I cannot believe that members of a clone will live together in sweet communion. Evolutionary humanists ignore the sin factor which inevitably spoils man-made Edens. "The crude evidence in human experience," says Dr. Paul Ramsey, "does not lend unequivocal support to the expectation that 'intimate communication' would be increased." To the contrary, Ramsey thinks that "animosity in personal relations" might be heightened.[2]

Ramsey further indicates that the struggle for selfhood and identity in a clone could be intense. "Growing up as a twin is difficult enough. . . . Who then would want to be the son or daughter of his twin? To mix the parental and the twin relation might well be psychologically disastrous for the young."[3] Personal and psychological independence, Ramsey projects, might be impossible to achieve for children who are the exact copy of their one parent.

6. *Cloning can provide a genotype of one's spouse, living or dead, of a deceased parent, or of some other departed loved one.*

Advocates of cloning suggest that couples unable to have children together might prefer a cloned child to one by artificial insemination since a third person would not be involved. Here, too, there could be an identity problem, unless the child was never told of his true parentage. Even so, the amazing likeness between the child and one of his parents would be obvious when the child looked in the mirror.

Cloning to preserve the genotype of a departed loved one will require that the necessary cells be removed before death. Biologists can already keep cells alive in a laboratory culture for an indefinite time. If and when human cloning is accomplished, such "restoration" of the dead will be possible. Dead men can now sire children through artificial insemination using sperm donated and frozen while they were alive.

Such a possibility raises a hard question. Whose child will the offspring be? The scientist who does the cloning, the relative who pays for it, or the deceased donor?

7. *Cloning can provide a form of immortality for donors.*

If so, it will be only for those who can afford to pay for the cloning.

But let's suppose cloning for this purpose is available to whoever desires it. We've all known parents who seek to relive their youth through a son or daughter. This can create serious psychological problems in the offspring. No youth should be burdened to live the life of a selfish parent.

Furthermore, there is absolutely no evidence that human consciousness can be transferred to a clonal descendant. The genes will be substantially the same, but the memory and thought patterns of the parent will not be reimplanted. This is just another vain hope of resurrection held by unbelievers. Christ alone is the resurrection and the life.

8. *Cloning can determine the sex of future children.*

Aristotle recommended that Greeks have intercourse in the north wind if they wanted males, in the south wind if they desired females. Women in the Middle Ages who wanted boys were told to drink a mixture of strong wine and lion's blood. Men in some European countries wore their boots to bed when they wanted to conceive a boy

Medical science today can give concrete help to improve the chances of producing the desired sex. Dr. Shettles was again one of the pioneers in this area.

It has been known for many years that if the sperm which fertilizes the egg carries an X chromosome, the product will be a girl; if a Y, the child will be a boy. The problem was that no one had been able to distinguish between androsperm (male sperm) and gynosperm (female sperm). Dr. Shettles examined some living sperm under a new phase-contrast microscope which revealed details ordinary microscopes had missed. He concluded that the male sperm were round-headed and the female oval-shaped, and that the males outnumbered the females in the ejaculate. Further experiments led him to believe that the larger number of androsperm was nature's way of compensating for their low resistance to acid secretions in the female vagina. He believed if the sperm could break through this acid barrier, they would have a better chance than female sperm of surviving in the alkaline environment within the cervix and uterus. From his observations he compiled a list of procedures which he advised couples to follow in sex-selection of children. In one case study nineteen of twenty-two couples who wanted girls were successful. In another test, twenty-three of twenty-six couples wanting boys got their hearts' desire.

Then while researching scientific and historic literature on sex-selection, Dr. Shettles found that Orthodox Jews produce more male offspring than does the general population. In the Jewish Talmud he found that rabbis had proposed one of the procedures which he had devised in his lab.

Dr. Shettles' sex selection techniques are not accepted by all doctors. He does not guarantee that the techniques will always work but insists that they will increase chances of conceiving a child of the desired sex.

The new prenatal test known as amniocentesis can now be used to identify the sex of a child before birth. The test is designed to identify certain genetic defects. However, doctors are faced with more and more parents demanding the test for purposes of sex selection. If the child is not of the sex desired, the parents request an abortion. Such callousness is a mockery of God's gift of life. More than that, abortion for this purpose is, in my view, first-degree murder. It lowers biological parents below the level of animals who would not reject their own flesh and blood.

With cloning there is no need to follow complicated procedures in intercourse or to have the amniocentesis test. The cloning of a man will always produce a male and of a woman will always result in a female.

9. *Cloning can increase scientific knowledge about human reproduction.*

Hitler's scientists increased their knowledge by studying human brains taken from the skulls of undesirables killed in pruning the race. This did not justify the murder of these innocent people. Unlawful and inhumane experimentation on other humans has occurred, even in the United States. I don't doubt that some American scientists have worked with aborted, live human fetuses. Their goal may be to learn more about genetic disease, but the work is still wrong—shockingly, appallingly wrong—and a gross violation of the rights of the unborn.

I have already said that a cloned offspring, prenatal or postnatal, will be a person. He or she should not be treated like a laboratory animal.

These are the nine most often mentioned scientific justifications for cloning humans. Cloning advocates differ over which they consider most important. Some scientists who favor cloning reject certain of the more trivial possible uses, such as cloning to determine the sex of the child.

A few scientists mention more remote reasons, such as cloning legless astronauts to take long journeys into space in cramped space capsules. This is more in the realm of space fiction than the cloning I discuss in this book. I am restricting my material to what I believe is possible within the next few years. We need to prepare answers now for the arguments being given in favor of this proposed drastic change in human reproduction.

CLONING
IS NOT
GOD'S
WAY

In the previous chapter I sought to expose the fallacies in the most common reasons put forth in favor of human cloning. Now I want to speak from a Christian perspective. The Christian world view is markedly different from that of the secularist who does not accept the idea of God or a purposeful, orderly universe. The secularist may call himself a humanist. However, if he is true to his philosophy, he will be more of a mechanist who sees humans in terms of function rather than of being. He will hold society or the common good in greater reverence than the individual. He will make individual aspirations and goals subservient to the good of the species as he sees it.

Christians begin with God, not man. We believe God has revealed his will for man in the Bible, with the foundations laid in Genesis. All that we see in nature and discover through human experience and scientific experimentation merely furthers our understanding of God's revelation in the Bible.

I'll give you a couple of examples. The Hebrews of Moses' day certainly didn't understand the laws of genetics. They had never studied cells under an electron microscope. They no more understood the relationship between recessive and dominant genes than they did the functions of electrons and protons in the atom.

God told them they were not to marry close blood relatives (Lev. 18:6-16). He didn't tell them why. The Bible is not a textbook of genetics or any other science, yet we know that the levitical law against incest makes good genetic sense. One who marries a close blood relative risks "stacking" bad genes which may result in retarded or diseased offspring.

The prohibition against adultery was also plain enough to God's ancient people. They didn't know, as we do, that the Seventh Commandment makes good sociological and psychological sense. They didn't have the statistical techniques and data processing equipment which we can use today to prove the wisdom of this scriptural prohibition.

These and other laws relating to marriage and family were given thousands of years ago. Yet they're as up to date now as when they were given.

Of course, God could have made it impossible for man to commit incest or adultery. But he chose to leave us options and to spell out the consequences of the wrong choice.

If human cloning is possible within the biological mechanisms which God has created, and I think it is, then short of the end of the world, God will not stop it from happening. If a human cell nucleus with forty-six chromosomes can be inserted into an enucleated egg cell and chemically triggered to start dividing and growing, then the resulting embryo will develop and grow into a fetus just as in ordinary reproduction. This will not be creating new human life, but simply reproducing life in a novel way. If it happens, I will not stop believing in God or the Bible. I will view it as just another attempt by man—albeit one with potentially grave consequences—to circumvent the plan of God for the perpetuation of the human race.

Human cloning will not, however, be wrong because it

is a new achievement in science. We Christians have too often been guilty of condemning a scientific development simply because it is new.

Such an incident played a role in the life of the second President of the United States. As a young man preparing for the ministry in Braintree, Massachusetts, John Adams became upset over condemnation of smallpox vaccination. Leading Congregational clergymen were calling smallpox a divine judgment for wickedness. It was wrong, they said, to interfere with the carrying out of God's judgment. For this reason and due to other wrangles within his church, Adams switched to law.

I want to say it loud and clear—neither genetics nor any other division of science is inherently an enemy of the Bible and Christian faith. Science is neutral, amoral. It can be used for good or evil.

Indeed, the mandate for honorable scientific research is found in the Genesis command to "subdue" and "rule over the creation." Says Donald MacKay, a highly renowned English brain physiologist:

> The key to the whole problem of the relation of science to the Christian faith, is that God, and God's activity, come in not only as extras here and there but everywhere. If God is active in any part of the physical world, He is in all. If divine activity means anything, then *all* the events of what we call the physical world are dependent on that activity.[1]

MacKay defines the role of scientists as "mapmakers in God's world . . . It is not just that we have permission, but that we have some obligation to get out there and map it so that the people who come after us can use the map. Mapmaking is wrong," declares MacKay when it "involves an infringement of the moral law. One should not become an adulterer in order to study the psychology of adultery. Apart from moral considerations, I know of no biblical guidance that prohibits any particular area of research."[2]

However, Dr. MacKay goes on to say, "Cloning [humans], as such, does not seem to me to transgress any biblical injunction."[3] The English scientist, who speaks frequently to Christian groups, thinks clonal offspring will be no different, in terms of human dignity, from identical twins. I disagree. There will be a great deal of difference. The birth of identical twins will continue to be a pretty rare occurrence. How many do you know? Once cloning is perfected, we may be faced with thousands, even millions of such humans. Given the right circumstances, I can visualize a thousand identical clonal twins in one city.

MacKay qualifies his endorsement of cloning by saying it might not be desirable to replenish the earth by reducing the variety of genetic types, as would happen in cloning. "To mess about with such a delicately balanced system without good reason could be disastrously irresponsible. It would also be very difficult to ensure proper psychological development in cloned children deprived of a normal family background."[4]

Here, in my judgment, Dr. MacKay has come close to the crucial biblical objection against cloning: the lack of a normal family background. God knew that children would need two parents, of complementary yet different sexual identities. "In the image of God He created him: male and female He created them. And God blessed them; and God said to them, 'Be fruitful and multiply' " (Gen. 1:27, 28). Clearly God intended for society to be built on the two-parent family.

We're now suffering from an epidemic of divorces and marital breakups. Single-parent families comprise a large segment of American society. Eight million U.S. households in 1978 were headed by women, a 44 percent increase since 1970. In many inner-city public housing projects, over 90 percent of the families are without a father. The crime rate in these projects is staggering. Willful, sinful rebellion is the root factor, but this is encouraged by poverty, unemployment, and the absence of paternal guidance and masculine role models. I don't doubt that the single largest cause of the upsurge in crime is the breakdown of the family structure. Hear what Dr. Leanor Turoff Glueck, a criminologist researcher at Harvard University, has to say:

> Every child has a right to parents who love each other, who want him and love him, who are deeply concerned for his welfare, who will give him supervision so he will not be so completely left to his own resources that he will make serious mistakes . . . Unless and until there is a concerted effort to preserve good families and reconstruct those that are not good, we are going to see crime penetrate every social level.[5]

God could have created mankind as a single sex. He could have created us as two or more sexes, each under one parent, to be perpetuated by cloning. But he didn't create just Adam or just Eve; or Adam as the head of one kingdom and Eve as the head of another. He made Adam and Eve to be partners in reproduction and the care of children, and to set the pattern for future generations. "For this cause a man . . . shall cleave to his wife; and they shall become one flesh" (Gen. 2:24).

In God's marriage mathematics, one plus one usually equals one, although it can also produce two, three, or more (multiple births). Still, each child has his own identity, even when he is an "identical" twin. The two parents still see their expression of love in each individual child.

Most families have a mixture of males and females. The males are outnumbered in my family three to two. Even in units in which the children are all girls or all boys, there is still one parent of the opposite sex, unless death or divorce has intervened.

The members of a clone necessarily must be all of one sex. A female clone can only have a female parent; a male clone only a parent of his sex.

Because of the difficulties inherent in a single-sex family, I do not think clonal children will be raised by their parent. I think they will be turned over to nurseries for raising. Sociological studies of children raised in nurseries are not encouraging.

One of the classic experiments was conducted by Dr. Rene Spitz of New York. He surveyed children in two similar institutions, which differed only in the type of personal care offered. "Nursery" babies were cared for by their mothers. "Foundlinghome" infants were raised by a few overworked nursing personnel. A guide called Development Quotient (DQ) was used to compare the development of the two groups. This included facility in control of bodily functions, muscle coordination, social relations, and general intelligence. "Nursery" babies started with a DQ of 101.5 and rose in one year to 105. "Foundlinghome" babies began with 124 DQ and declined in two years to 45. In two years 37 percent of "Foundlinghome" children died, but in five years not a single child was lost at "Nursery."

Members of a clone will probably be more like the babies in "Foundlinghome" than in "Nursery."

I haven't mentioned the importance of role models and relationships in a family. Two parents are needed for the development of good sexual identities in children. Many, if not most, of our growing homosexual population come from homes devoid of healthy role relationships with parents. It is entirely possible that large numbers of clonal children will turn toward homosexuality or other sexual perversions.

I've already said that the family has been battered in recent years by destructive influences. Radicals say the family is dying and will soon be obsolete. They see cloning and other developments in biological science as encouraging this trend.

Those who hold this view also claim that when culture is stripped away there are only biological differences between the sexes. Our ideas of femininity and masculinity, they say, are environmentally conditioned.

A man or woman ought to be able to have any job for which he or she is qualified. I believe in equal pay for equal work. However, there are many ways, besides biological, in which the sexes will always differ.

Perhaps you've heard of the *yin* and *yang* principle. *Yang* is the masculine or positive force, *yin* the female or negative force. Oriental Taoists believe the interplay of these forces occurs in every activity of the universe. At one time in one place *yang* may be stronger, at another time and location *yin* may dominate. This is a beautiful illustration of the marriage relationship, reproduction, and the family interactions that support the continuum of life. Human cloning, if done on a large scale, will probably upset this God-ordained balance of gender. We will all be the losers for it.

I agree with Paul Ramsey that the cloners will be no more than technicians in animal husbandry. In their attempt to improve humanity they will lower mankind. Ramsey warns:

> To put radically asunder what nature and nature's God joined together in parenthood when he made love procreative, to disregard the foundation of the covenant of marriage and the covenant of parenthood in the reality that makes for at least minimally loving procreation, to attempt to soar so high above an eminently human parenthood, is inevitably to fall far below—into a vast technological alienation of man. Limitless dominion over procreation means the

boundless servility of man-womanhood. The conquest of evolution by setting sexual love and procreation radically asunder entails depersonalization in the extreme. The entire rationalization of procreation—its replacement by replication—can only mean the abolition of man's embodied personhood.[6]

Another major objection to human cloning from a Christian and humane point of view involves the destruction of clonal life in experimentation. Such wanton destruction of unborn humans, whether in embryo or fetus form, must be opposed. This is not just a conservative Christian position. It is also strongly held by many other groups in our society who believe that the right of the unborn to live should take precedence over the curiosity of scientists to experiment.

Some time ago, when human cloning was just beginning to be written about in newspapers, Dr. Leon Kass responded to a column by Dr. Joshua Lederberg promoting cloning in the *Washington Post*. In a letter to the editor, Kass asked what would happen to mishaps in cloning. "Who will or should care for 'it' and what rights will 'it' have?" he wanted to know. "Will the programmed reproduction of man not in fact dehumanize him?"

Kass further wondered if it made "sense to say that each person has a right not to be deliberately denied a unique genotype. Is one inherently injured by having been made the copy of another human being, regardless of which human being?"[7] A thorn in the side of scientists who live by consequentialist ethics, Kass holds that even to attempt cloning is unethical because it will place in jeopardy the clonal offspring, whose permission cannot be obtained.

The decline of reverence toward fetal life since the Supreme Court voted to allow abortion on demand is already dismaying. When I was in western Canada on a lecture tour last year I discovered that this malaise of the spirit is not confined to our borders.

One night I was interviewed by a reporter in Saskatoon, Saskatchewan. He mentioned that his sister was a neo-natal nurse, one who works with premature infants and babies with birth defects. "How does she feel about abortion?" I asked. "Oh, she's dead-set against it," he replied. "What really bothers her is that in one wing of the hospital they're killing babies older than the ones she's helping to save in her wing."

Over a million babies will be aborted this year in the United States alone—forty million worldwide. Population explosion or not, I think this is nothing short of mass murder of those who cannot protect themselves.

Cloning, accompanied by the destruction of laboratory mistakes, will only add to the callousness which threatens us all. I am totally against this scheme to propagate children by asexual reproduction. I don't see an embryo or fetus as just a pulsating blob of tissue. This precious life, whether in a woman's womb or in a laboratory culture, is the fruit of God's own creative processes, a human in miniature, and at the very least a life in the making. The right to live should be preeminent.

Human life is too precious for experimentation by would-be cloners, whether they be mere graduate students or Nobel Prize winners. In the blind rejection of the Creator's plan, they may honestly feel that cloning will benefit the human race. No matter. They are not wiser than God. They should not be allowed to carry on their experiments when human life is at stake. We should rise up and say so now before C-Day—before a team of scientists stands up before the world and grandly announces, "We have cloned a man."

THE TEST-TUBE BABY

While a biology professor was away from his lab, his students decided to have some fun. They glued together an assortment of body parts from various bugs—the legs of a grasshopper, the wings of a butterfly, the torso of a black beetle, and the head of a caterpillar—and placed their creation on the teacher's desk.

When the professor returned, a student asked, "Prof, how would you classify this, uh, specimen?"

The teacher picked it up gingerly and turned it over, pretending to inspect it closely. Then he announced with great solemnity, "Ladies and gentlemen, this is a humbug!"

You may still believe that cloning a human is all humbug—foolish speculations by a few scientists trying to grab public attention.

You can't say that about the so-called test-tube baby. Baby Louise Brown was born by Caesarean section at Oldham General Hospital in northern England, July 18, 1978. She is the first child known to have been conceived in a laboratory culture dish. Gynecologist Patrick Steptoe, then a staff doctor in Britain's National Health Service, and Cambridge University physiologist Robert Edwards, share the honors of bringing about this amazing scientific achievement.

The London *Daily Mail* bought pictures and exclusive rights to the Louise Brown story from the parents. The publicity extravaganza took the family to Disney World in the U.S., and to Japan, where the crowds gazed at the blonde, blue-eyed baby in amazement. Her parents, truck driver Gilbert Brown and his wife Lesley, received enough money to move from their small house in the British port city of Bristol to a large country home.

No birth of a princess has ever excited so much attention. The event was headlined around the world. Many reporters compared the scientific breakthrough to the splitting of the atom and predicted it would have far greater consequences.

Biologists were not taken by surprise. Many had been predicting that a test-tube baby would soon be born. They were not all agreed that it would be a good thing. Dr. James Watson, who had shared a Nobel Prize with two colleagues for unraveling the structure of DNA, the chemical blueprint of life, had warned a U.S. congressional subcommittee years before that such a birth would occur, predicting, "all hell will break loose, politically and morally, all over the world."

That hardly happened. After the headlines and a flurry of TV specials, the excitement cooled. Most doctors wondered what all the fanfare had been about in the first place. They saw no moral problem so long as it was a husband's sperm and a wife's egg that had been mated in the culture dish. Human life had not been created in the laboratory. Steptoe and Edwards had only built a detour around the blocked channels for the sperm and egg to achieve fertilization. This, they said, was a godsend to the 1 percent of women who cannot conceive because of blocked fallopian tubes.

Each of these assessments falls short of the mark. The test-tube baby has not by itself ushered us into a world of mechanical reproduction. But when considered in the larger context of the biorevolution, it may have far-reaching implications. For one thing, it brings human cloning a giant step closer.

To understand what Steptoe and Edwards accomplished and the potential effects of their triumph, we should first review the early stages of the reproductive process. The female reproductive organs are shaped something like a T with the small twin ovaries located underneath the curving arms of the fallopian tubes. The sperm enter the uterus and swim "up" to the point where the tubes branch off at the top of the T. If it is the right time in the menstrual cycle a ripe ovum (egg cell) breaks off from an ovary and meets the sperm in one of the tubes. Conception occurs when the egg and a sperm join. The fertilized egg, now called a *zygote,* travels down this tube and burrows itself in the lining of the uterus where a nesting place has already been prepared. Snug and nourished by the mother's bloodstream (barring damage by accident or disease), the embryo will develop until ready for birth.

Blockage of the fallopian tubes is like having the two lanes of a divided tunnel barricaded. The crucial connection between sperm and egg cannot be made in either lane. What Steptoe and Edwards did was to remove the sperm and egg and initiate the crucial fertilization in the laboratory.

Steptoe and Edwards built upon the achievements of other scientists. In the 1940s, Dr. John Rock of Harvard, "father" of the Pill, simulated the chemical environment of a woman's fallopian tubes in a laboratory culture dish. He succeeded in fertilizing eggs (taken from female cancer patients) with sperm and brought several embryos to a multi-cell stage.

In 1950 Dr. Landrum Shettles kept a laboratory-conceived embryo alive for six days outside the womb.

Eleven years later, Dr. Daniele Petrucci of the University of Bologna kept an embryo alive for twenty days, until it became deformed. The following year, Dr. Petrucci nourished an embryo for fifty-seven days during which he recorded fetal heartbeat on an EKG machine.

Communist Chinese newspapers called Petrucci's experiments happy news for women, providing hope for the development of methods of reproduction which would not burden working women with childbirth. In Italy the news of what he had done caused an uproar. Under a blast of criticism from the Vatican, Petrucci moved to the Soviet Union and joined a team of Russian scientists at work on an artificial womb.

In 1971 Dr. Landrum Shettles at Columbia Presbyterian Hospital in New York City fertilized an egg from one woman, grew it to the sixteen-cell blastocyst stage, and implanted it into the uterus of a second woman scheduled for a hysterectomy. The embryo grew to several hundred cells and was then removed during surgery. A Florida dentist and his wife heard about Dr. Shettles work and asked if he would help them have a baby.

Shettles saw the woman, Delores Del Zio, as an ideal candidate. She was healthy, young, and her fallopian tubes were blocked. Assisted by Dr. William J. Sweeney, Shettles fertilized Mrs. Del Zio's egg with her husband's sperm and achieved conception in the lab. Two days before the embryo was due to be implanted into the woman's womb, the developing life was allegedly confiscated by Dr. Raymond Van de Wiele, Shettles' superior at Columbia. Van de Wiele later told David Rorvik that Shettles had never obtained clearance for the "immoral" and "unethical" project. Shettles held that permission was not necessary since Mrs. Del Zio was Dr. Sweeney's patient and the actual implant was to have been done at another hospital. Shettles added that Van de Wiele had "become emotional" and ordered him out of the hospital.[1]

The Del Zios sued for $1.5 million, contending that the hospital and Dr. Van de Wiele had deprived them of their property and that substantial damage had been done to

Mrs. Del Zio's psyche. A jury agreed, but awarded only $50,000. The legal hassle between the two contending doctors never reached court. Dr. Shettles resigned "voluntarily."

Meanwhile, English scientists were experiencing no legal difficulties in attempts to produce a test-tube baby. Dr. Douglas Bevis made his surprise announcement that he had presided over the birth of three test-tube babies by the IVF method. Bevis freely admitted previous failures resulting in the death of embryos. "So many have been attempted," said Bevis, "that by the law of averages some have come through."[2]

Bevis' claim was not accepted by his colleagues because he produced no proof. Newspaper stories of his and Shettles' work did arouse opposition from many church leaders and some scientists. Some of the churchmen quoted German theologian Dietrich Bonhoeffer, a martyr of Nazi Germany during World War II, who argued that an embryo's existence is itself evidence of God's intention to create a human being, therefore the embryo's right to life is divinely bestowed and any deliberate deprivation of it is nothing short of murder.[3]

Criticism over destruction of embryos in test-tube baby experiments resulted in the U.S. government banning new grants for *in vitro* fertilization. Denied federal money, IVF work in the U.S. was virtually stopped.

The English doctors, Steptoe and Edwards, had continued to push ahead. When the Browns came to them for help in 1977, they were ready. In the successful experiment, Steptoe, the surgeon, made a small incision just below Lesley Brown's navel. He inserted tiny forceps, grasped the ovaries, and extracted as many tiny eggs as he could grasp, about fifteen. He then transferred these eggs to a culture dish containing a special mix of hormones and other nutrients which he had concocted to simulate the womb environment.

Edwards now removed and washed the eggs in a chemical solution before mixing them with 50,000 or so of Mr. Brown's sperm cells. Periodically, he and Steptoe examined the eggs and sperm under a microscope for evidence of fertilization. When this occurred, they transferred the new embryo to still another dish. Within four days it grew to thirty-two cells.

Mrs. Brown was brought back to the operating room where Steptoe used microscopic controls to insert the embryo into her womb. Slightly less than nine months later Baby Louise Brown was born.

Steptoe came to the United States and presented a full account to the American Fertility Society at its meeting in San Francisco. He revealed that he and Edwards had made thirty efforts to implant embryos before succeeding. The failures did not appear to bother the 1,200 doctors and biomedical specialists present. They gave Steptoe a thunderous ovation.

Steptoe and Edwards' success stirred U.S. scientists to ask that the ban on federal grants for *in vitro* research be lifted. The Ethics Advisory Board of HEW held hearings in eleven cities. Those who favored abortion on demand tended to support test-tube baby research. They generally took the position of Joseph Fletcher that "a fetus is not a moral or personal being since it lacks freedom, self-determination, rationality, the ability to choose either means or ends, and knowledge of its circumstances."[4] Those who wanted the spigot of federal money to remain turned off pleaded the unborn's right to life. Princeton's Paul Ramsey also feared that physical or psychological damage could be inflicted on IVF children.

Test-tube baby research has continued abroad. A second IVF baby has been born in India and a third in Scotland. The technique involved in the conception and implantation of the baby born in India differed from the Steptoe-Edwards method in two ways. First, several eggs were fertilized for implantation, giving better chances of success if a number of embryos were implanted. Second, the fertilized eggs were kept frozen until implantation at the time in the mother's menstrual cycle when chances of acceptance were considered optimum.

Few U.S. doctors and scientists seem to be alarmed. They argue that laboratory conception and implantation of an embryo is no different in principle from taking fertility drugs or having a baby by Caesarean section. The question of whether man should interfere with the natural processes of human reproduction, they add, has already been decided. Norfolk General Hospital has already applied for and been granted permission by the Virginia health commissioner to establish the first test-tube baby laboratory in the U.S. Several other U.S. hospitals are planning to open test-tube baby centers as soon as they can obtain official clearance.

Biological scientists generally see *in vitro* fertilization as another step in the effort to improve the quality of human life. Laboratory conception and embryo growth will enable researchers to study genetic defects more closely. More might be learned about Down's syndrome, for example.

It may be possible, they say, to detect abnormalities by extracting a few cells from an early embryo for study of abnormalities. The embryo could be frozen while the cells are multiplying in a culture and being checked for defects. If the cells prove healthy, the embryo can then be thawed and implanted within the uterus. If the embryo is defective, it may still be possible to correct malformed or malfunctioning genes. If the disease is deemed incurable, then the embryo can be used for experimental purposes, for example, testing to see how much radiation the cells can endure or determining what drugs are harmful and in what dosage.

Anti-abortionists say this is no different from ordinary abortion. The researchers reply that a large percentage of diseased embryos spontaneously abort anyway. Anti-abortionists retort that this cannot be determined until the embryo is given a chance to fight for life.

Those who believe that unborn humans have rights also object to this practice on the basis that it is unlawful to experiment on live members of our species. It cannot be justified, they say, on the basis of potential benefit for other people or the race in general.

It is now unlawful to experiment on live humans without consent. Before World War II this was only a consensus in many countries. It was felt that medical science would abide by the unwritten rules. Then the horrors of Nazi concentration camps and experimental hospitals were revealed. The world learned that German medical personnel had been instructed to go through their wards and select the least fit, to use for experiments. Some hospital workers picked out known troublemakers and other patients they disliked. German researchers did not care who was chosen so long as they got their daily supply of fresh human organs.

Disclosures of what happened in Nazi Germany resulted in the Nuremburg Code of 1949, the first code of medical ethics for research on humans. This was adopted by the World Medical Association at its first 1964 meeting in Helsinki, Finland. The code called for:

—Experimentation to "conform to the moral and scientific principles that justify medical research" and to be "based on laboratory and animal experiments or other scientifically established facts."

—"Every clinical research project" to "be preceded by careful assessment of the inherent risks in comparison to forseeable benefits to the subject or to others."

—The doctor "to remain the protector of the life and health of that person on whom clinical research is being carried out."

—The doctor to explain "the nature, the purpose and the risk of clinical research" to the subject.

—The doctor to "obtain . . . in writing the consent of the subject."[5]

By 1964 the Nazi atrocities were only a bad memory for most people. Trust in the benevolence of doctors climbed. Then a U.S. researcher in anesthesiology, Henry Beecher, unveiled a bombshell report, charging that as many as 12 percent of the experiments being conducted in the U.S. by qualified medical researchers were unethical.

Beecher noted that twenty-three charity patients had died after being treated with experimental drugs in an effort to find a better treatment for typhoid fever. Had they received proper care, Beecher said, they would have been expected to survive. He mentioned another case in which a researcher, identified only as Dr. S., injected live cancer cells into twenty-two human subjects, telling them only that they would be receiving "some cells." Dr. S. was later brought before the New York State Board of Regents, the licensing agency for physicians in New York.

An associate of the accused doctor said they had not told the patients the injected cells were cancerous because "we did not wish to stir up any unnecessary anxiety, disturbances, or phobias." A reporter cornered Dr. S. afterward and asked if he would accept an injection of cancer cells into his own bloodstream. He would not because "there are relatively few trained cancer researchers, and it seems stupid to take even that little risk." In his incriminating report, Beecher said Dr. S.'s research colleagues had *not* disapproved of his work, indeed the American Association for Cancer Research elected him as its vice-president in 1968, and president in 1969.[6]

These and other examples of unethical human experimentation are included in *Ethical Decisions in Medicine* by Howard Brody, a standard text in medical ethics now being taught in some medical schools. Three ethical theories are set forth in this text: 1. Consequentialist ethics "in which an action is judged to be morally right or wrong by judging the consequence of the action"; 2. Utilitarian ethics, a special kind of consequentialist ethics, which states that the "ultimate principle against which consequences are to be judged is the general happiness of all people concerned, or the greatest net balance of good over evil"; 3. Deontological ethics, which insists "that there are rules or principles of action which have moral validity independent of the consequences of individual actions, and that one must act in accordance with these rules or principles." The latter theory, according to author Brody, is usually held by persons with strong religious beliefs.

Dr. Brody clearly lines up with consequentialist ethics, and suggests that medical practitioners and researchers consider the "risk-benefit ratio." "The potential benefits to be gained, both by the individual and by society as a whole, must be weighed against the risks of untoward side effects to the individual," he declares.[7]

Consequentialist ethics asks the researcher or his employer to judge whether a specific experimentation is right or wrong. I'll admit that we all make decisions every day based on what we think is best. There are simply not enough clear guidelines to cover everything in our modern technological world. But should scientific researchers or anyone else have the authority to decide who shall live and who shall die through experimentation on unborn human life?

This problem, of course, relates to an issue we've already raised: Is an embryo a person? Does it have a right to life? If we answer yes, then there is the matter of consent. Obviously, the embryo's consent cannot be obtained. Then who should consent? The parents? The researcher? Is the good of the race—a cure for cancer, for example— sufficient reason to give consent?

Ramsey thinks the health of the individual patient, prenatal or postnatal, must be put before any potential benefits for that "non-patient" (Ramsey's term), the human race. This seems to be a reasonable guideline.

There's another implication of IVF which we should consider. This is the possibility of a host mother bearing a child in place of the child's biological mother.

The host mother concept was demonstrated in animals long before the first test-tube human baby was born. Back in 1962, English scientists extracted two fertilized eggs from sheep, tucked them in the warm oviduct of a live rabbit, and shipped the hare to South Africa. Scientists there removed the eggs and implanted them in two ewes which gave birth to healthy lambs. Here three "mothers" were involved, a biological mother, a transfer mother, and a birthing mother.

Scientists in the United States soon did their English colleagues one better. They gave a prize cow fertility hormones to produce a large supply of ripe eggs. They fertilized these eggs with sperm cells from a select bull, then took the resulting embryos and implanted them into the uteri of ordinary cows. Their reward was a whole herd of calves with superior heredity.

The same could be done with a "prize" woman. Give her fertility hormones. "Breed" her to a prize man. Extract the harvest of fertilized eggs. (A Naples, Italy, woman, who had taken fertility drugs, gave birth to octuplets in 1979; only two survive at this time. Previously she had delivered sextuplets, but none survived.) Implant the fertilized eggs in the uteri of a number of healthy women. The result will be a "herd" of people. Terms such as "breed," "prize," and "herd" fit perfectly into the context of this sort of "animal farm." With a totalitarian government on the order of Nazism or Communism and a few unscrupulous scientists, a whole chain of state-run farms could soon be producing humans in this fashion. *Brave New World* may not be science *fiction* much longer.

We're not done with this monster which biological scientists have the potential to make. Remember Dr.

Petrucci, the Italian scientist who was driven out of Italy by the Vatican? Using his knowledge, Russian researchers are said to have about 250 human fetuses growing in artificial wombs at the Institute of Experimental Biology in Moscow. The Soviets have clamped a tight news lid on their experiments and are not permitting pictures or interviews by western journalists. It is possible they have produced some malformed babies which they don't want the world to know about.

Artificial wombs that carry babies to term will probably be available in ten to twenty years, if not before. Science is moving closer from both ends of the gestation process. More fully developed embryos are being grown in the lab and special environments for "preemies" with hyaline membrane disease are already in the hospital nursery. When the technology for growing embryos and saving preemies meets, we'll have artificial wombs. Then a woman won't have to hire a host mother. She can donate her egg and let a machine do all the work.

Childbirth for the rich will be painless. They can leave sperm and eggs at a lab, go about their regular business, and drop by the lab or hospital in nine months to collect their baby. In the meantime, if they should decide raising a child would be too much bother, they can offer it for sale. Perhaps they could put an ad in the classifieds saying something like: "Healthy, male Caucasian fetus now developing in Broadway Laboratory. Parents of good stock. Pedigrees available for inspection. Make offer."

Dr. Leon Kass calls the development of the test-tube baby "dehumanization" of the birth process. "Quality control of the product" he says, is not worth "the depersonalization of the process." He asks, "Is there not wisdom in the mystery of nature that joins the pleasure of sex, the communication of love, and the desire for children in the very activity by which we continue the chain of human existence?"[8]

78

7

BABIES
WITH TWO
"FATHERS"

The science of producing test-tube babies is extremely new. There are only three of these babies living at this writing. But we need to consider the effect of this procedure on society when it becomes more common.

It's too easy to say, "We'll just have to wait and see." I think we can predict, at least to some extent, what will happen by looking at a parallel procedure in human reproduction which has been available for many years. I'm speaking of artificial insemination as it is practiced when the husband is sterile. In the procedure, called AID (artificial insemination by donor), the physician simply inseminates another man's fertile sperm directly into the woman's uterus and nature takes over from there.

Because of the secrecy involved, no exact tabulation is available on the number of AID offspring. Estimates within the U.S. medical profession run from 200,000 up to 500,000, with 5 to 20,000 new AID babies born each year.

IVF requires that the sperm and egg be removed and mated in a laboratory dish, then the resultant embryo is implanted in the woman's uterus. AID is done in a few minutes in the doctor's office. The physician schedules a woman's appointment for the time in her menstrual cycle when she is most likely to conceive. When his patient arrives he is ready with a specimen of sperm from a donor or a sperm bank. He merely inserts a portion of the sperm into her cervix with a tuberculin syringe, then tucks inside her vagina a tiny plastic cup containing the remainder. The doctor instructs her not to swim or douche during the next six to eight hours, then to remove the cup.

If the first attempt doesn't succeed, she returns about a month later for a second attempt. Most healthy women conceive after one or two inseminations with fresh sperm from a donor; frozen sperm usually requires three or four treatments.

I must digress here and explain that artificial insemination is also done with the husband's sperm when he cannot impregnate his wife through ordinary intercourse. He may be paralyzed, yet can still produce sperm. His sperm count may be too low. The physician will take several collections of sperm and inseminate the concentrate. There may be some chemical incompatibility between his sperm and the acid secretions in his wife's vagina. The doctor will take a sperm specimen and bypass the trouble spot by inseminating the sperm directly into the fallopian tubes.

Or the husband may wish to have his sperm frozen for later insemination. He may be going off to war or undertaking a dangerous job assignment. He may have to undergo surgery which will render him sterile. He may wish to have a vasectomy. By storing his sperm he can father a child at a future date of his and his wife's choosing.

The propriety of AIH (artificial insemination husband) is questioned by some Roman Catholic authorities who judge the rightness of the procedure by the method in which the sperm is obtained. They say the act is licit if the sperm is obtained during intercourse, but wrong if secured by masturbation. Protestant theologians generally do not object to AIH in any form.

Medical specialists tend universally to recommend AIH as a worthy means of conceiving children. Dr. S. J. Behrmann, former director of the Center for Research in Reproductive Biology at the University of Michigan, recalls helping a husband who was paralyzed and could not have normal relations with his wife. "A year or so later, I saw this man in his wheelchair holding his child. He was crying for joy."[1]

AID, as you might guess, is much more controversial. Some doctors who do AIH will not administer AID. I tend to agree that the minuses outweigh the pluses, yet cannot be too judgmental about couples who take this route after all else has failed. I suggest you withhold judgment until we explore the facts about this sensitive procedure which raises so many legal, moral, and medical questions.

Let's look first at the history of AID. The practice began as an aspect of animal husbandry. One of the first accounts in ancient literature tells of an Arab sheik, living around 1000 A.D., who inseminated his enemy's sleek mares with sperm from sickly stallions. A few years later the inseminator won an easy military victory. Around 1780 an Italian physiologist, L. Spallanzani, inseminated a dog to produce a litter of three pups.

England claims the first human birth by artificial insemination. About 1800, Dr. John Hunter successfully inseminated the wife of a London linen merchant. The first American AID success was reported in 1866 when Dr. J. Marion Sims claimed that six of his women patients became mothers through donor sperm. He didn't name the women.

AID was a well kept secret within the medical fraternity before World War II. Today, when almost anything can be printed or shown on a screen, it is hard for us to imagine a time when words such as womb, ovary, sperm, oviduct, and sometimes even pregnancy were taboo in public newspapers. Bearing a child out of wedlock was a scandal then.

For fear of their reputation and also of prosecution, doctors didn't keep records of donors and recipients. The legal husband's name always appeared on the birth certificate, even though he was not the biological father. We will never know how many family lines are broken by artificial insemination.

Cattlemen did keep records. In the 1930s the first AID calves were born in Wisconsin. Half of all American dairy cows are now sired by sperm taken from prize bulls. Dairy farmers employing AID have increased milk yield up to 65 percent by selective breeding.

Cross-breeding of animals is now common around the world. Peruvians, for example, have crossed llama males, which once provided clothing for Inca royalty, with alpaca females. The result is an animal with a fine grade of fleece.

Animal and human AID would not be possible on a large scale today without the development of deep freeze sperm banks. Again, the English take the honors in pioneering. In 1959, an English scientist, Dr. C. Polge, found that he could protect frozen sperm by adding a quantity of glycerol. Semen, containing sperm, is mixed in a twelve-to-one ratio with glycerol and kept in a liquid nitrogen freezer at -196 degrees centigrade. The sperm is thawed by merely exposing it to room temperature.

Large livestock breeders now routinely use sperm banks for improving their herds. Sperm from one bull may be used to produce a thousand calves in one year. Human AID practitioners in the U.S. serve their patients from eight central sperm banks and seven branch banks. The only reported accident has been the breakdown of a freezer resulting in the loss of hundreds of specimens. Thirty-five clients, including some single men who had had vasectomies, and cancer patients who had stored sperm while undergoing radiation treatments, have sued the bank owners.

Frozen sperm can be kept potent up to ten years, although samples more than two years old are seldom used. Some doctors disdain the sperm banks and will use only fresh semen which they claim gets better results. Drs.

Keith Smith and Emil Steinberger of Houston have recorded a conception rate in their patients of 61 percent with frozen sperm, compared to 73 percent for fresh ejaculate.

The going payment for a sperm specimen is thirty to fifty dollars per sample. Medical students, interns, and residents are preferred. They are young, easily available, usually need the money, and are considered more healthy than men in the general population.

Reputable AID physicians claim to counsel couples seeking AID. They want to know how badly they want a child and if they are mature enough to handle psychological problems that may come up. The doctor's fee is covered by Blue Shield in most states as a surgical procedure. AID is not now available to charity patients, except for voluntary experimental purposes. This has brought the complaint that AID is only for those who can afford it.

Whether they use an immediate donor or a sperm bank, AID doctors try to select sperm from men who match the husband's physical appearance and blood type. The idea is to make everyone, even the grandparents, believe that the baby is really the husband's child.

Many couples seek AID because of fear that their own union could produce a child with genetic disease. Until recently, it was presumed that all doctors ran careful hereditary checks on sperm donors. A University of Wisconsin survey of 379 AID physicians, the most comprehensive of its kind ever conducted, reveals that many do not. Only 12 percent of the physicians replying to the questionnaire said that they did chromosome tests on donors to prevent birth of a Down's syndrome child. Only 30 percent admitted checking for traits that might result in other serious genetic defects. Ninety-four percent said they would not use a carrier of Tay-Sachs disease as a donor, yet fewer than 1 percent tested for this affliction.

Over two-thirds of the doctors replying confessed that

they kept no files on donors. Some had used the same donor for several pregnancies; one for six was not unusual and one doctor reported the "fathering" of fifty children by a single donor. This raises the specter of half brothers and sisters innocently marrying one another, thus increasing their chances of bearing defective children. In 1974 an engaged couple who grew up in the same town cancelled marriage plans after learning from their family doctor that they were actually half-brother and half-sister.

Forty-seven physicians said they had inseminated single women, including lesbians, upon their request. Could a group of feminist extremists get control of a sperm bank? Well, in December, 1975, about thirty men, claiming to be for male supremacy, demonstrated outside Massachusetts General Hospital in Boston, demanding that sperm banks be closed because feminists were using them to have children outside of marriage.

The doctors who administer AID are caught in a bind. Some—the majority, I would hope—do not want to help unmarried women, particularly lesbians, to have children. But the tide of public opinion in this country is in favor of so-called equal rights. Therefore an AID doctor may not be able to refuse the request of any woman for insemination, so long as she can pay for it. If a woman can demand and get an abortion, I don't see how AID can be withheld from lesbians.

I would like to pause to mention a technique which, if perfected, would appeal to lesbians more than AID. Dr. Pierre Soupart of Vanderbilt University Medical School is developing a method of starting a new life with two eggs rather than an egg and a sperm. The offspring would have two biological mothers and would be a female. "It's the answer to a number of women's dreams," says Lucia Valeska, co-executive director of the National Gay Task Force. Each woman would donate an egg which would then be treated in such a way as to cause the eggs to fuse into one cell. That cell would be comparable to the zygote that results from the fusion of sperm and egg. After a few cell divisions in the laboratory, the growing embryo would be implanted in the uterus of one of the mothers. Dr. Soupart has carried out this procedure at least once with mice and has plans for further improvement of the technique. Lucia Valeska says, "It's very understandable that women who want to raise children together should want the children to be part of their biological makeup."[2]

Egg fusion is a radical departure from God's plan for human reproduction, and because it is not yet available, we can only speculate on the effects of its use. But AID is with us now, and the questions it raises are of immediate concern. In addition to the use of AID by lesbians, controversy centers on possible psychological and social problems in an AID family. Does having an AID child help or hinder a marriage? Dr. Behrmann, who formerly worked at the University of Michigan, knows of only one divorce among 600 recipients. Dr. Sheldon Payne of the Shelton Clinic in Los Angeles says he has studied AID couples over a span of thirty years and found their divorce rate to be one-fifth of the California average. He admits this may be because of the way couples are screened before being accepted for AID at his clinic. Having a child to stabilize a marriage, he says, is an immediate disqualifier.

On any practice as controversial as AID, you can usually find statistics and stories to support both sides, depending on who is doing the talking. I don't doubt that Drs. Behrmann and Payne are telling the truth about *their* patients. But what about the patients of doctors who are not so conscientious?

I won't make any judgments here, except to note that one medical researcher has discovered some bad results of AID. He is Dr. Laurence E. Karp, co-director of the Prenatal Diagnosis Center of the Department of Obstetrics and Gynecology, Division of Medical Sciences of the University of California at Los Angeles (UCLA), located at Harbor General Hospital in Torrance, California. The following examples are from his book *Genetic Engineering: Threat or Promise:*[3]

> —In two cases where AID was urged by doctors on couples, the husband of one and the wife of the other suffered serious mental depression afterwards.

> —A British physician gave up AID because he perceived that the practice caused husbands to feel inferior and the wives to have undesirable mental attitudes toward the anonymous donors.

> —The depression experienced by five husbands from not being able to father children was worsened by their wives having children by AID. The wives reportedly had sexual fantasies about the unknown donors and the doctors and rejected the babies prior to birth. After birth, all of the children suffered severe emotional damage.

These may represent only a small percentage of the thousands of couples who opt for AID, but they should make any couple or counselor think a bit before resorting to AID.

The legal problems also are troublesome. Only four states, California, Georgia, Oklahoma, and Kansas, have laws legitimatizing AID offspring, and then only if the husband has consented to the insemination. The Oklahoma law stipulates that a couple desiring AID must go before a county judge and sign consent. This is filed with the court and is similar to adoption. The AID child, when born, holds the same rights as if he were the couple's natural child. One other state, Arkansas, has a law protecting inheritance rights of AID offspring.

Results of the few court cases related to AID have been contradictory. The first North American legal battle occurred in 1921 and resulted in an Ontario court deciding that AID constituted adultery. The first U.S. case, occurring in Illinois in 1948, culminated in a different opinion. The judge said AID could not be adultery, then granted the husband a divorce because the wife had engaged in regular adulterous liaisons. The year before, a New York court also held that AID was not a form of adultery and ruled the child legitimate, provided the husband had consented to the procedure. "The child has been potentially adopted or semi-adopted by the defendant [the father]," the judge said. "He is entitled to the same rights as those acquired by a foster parent." The mother later moved to Oklahoma where she was granted a divorce and custody of her child.

In a 1954 divorce case heard by the Superior Court of Cook County, Illinois, the judge noted that an AID child had been born in lawful wedlock. The next year this court reversed itself and ruled that AID "with or without the consent of the husband, is contrary to public policy and good morals, and constitutes adultery on the part of the mother. A child so conceived is not a child born in wedlock and therefore is illegitimate."[4]

Few AID-related cases have come to court since 1954. None clear up the legal thicket in which AID participants are caught in most states, and the Supreme Court has never ruled on the subject.

Thousands of children have been fathered through AID, but the parentage of only a few is known. The usual custom is still for the husband to be declared the biological father. Sometimes the obstetrician knows the truth, yet falsifies the birth certificate to avoid trouble. Sometimes he doesn't know and innocently misrepresents the parentage. In many instances, a couple will go to a fertility specialist in another city. The woman receives donor insemination, returns home, and shows up at her doctor's office pregnant.

I can understand why the husband is declared the biological father in most situations and why doctors keep such poor records. The potential for trouble is great. A husband might decide to sue a donor for alienation of affection. An AID child might demand participation in his biological father's estate. Should the husband and wife die, relatives might demand that the AID child be disinherited. Any lawyer could suggest a dozen more possibilities for court action.

A host mother, who carries to term another's test-tube baby, will present the potential for similar problems. More court cases may result from the use of host mothers for the simple reason that it is easier to conceal a sperm donation than a pregnancy.

AID and host mothers also pose problems in the moral arena. Does the bringing of a third party—a sperm donor or a host mother—into marriage constitute adultery? One might argue that the use of a host mother does not constitute adultery, since she herself would not conceive the child. In any regard, there is as yet no actual case to consider. I'll go on to the moral question—is AID adultery? There has been a lot of discussion on this in church circles.

The Catholic position is that AID is adultery. Fr. Francis Filas, Professor of Theology at Loyola University in Chicago, bases the following view on "principles of official teaching":

> Adultery is the violation of the marriage bond which is oriented to new life and in which husband and wife have a right to each other's life-giving powers. The husband's right to his wife's procreative powers is a gift from God which he cannot give away. He can no more tell his wife to receive semen from a donor than he can tell her to have intercourse with another man.[5]

Protestant denominations are not certain that AID is adultery. The United Presbyterian Church endorsed a report in 1962 that accepted AID with the qualification that doctors should be sure of the "intelligence and emotional stability" of couples before advising this "radical social procedure." The Presbyterians refused to call AID adultery. This, they said, would give adultery "a meaning it does not have in the New Testament."

In 1970 the Lutheran Church of America declined to interpret AID as an act of adultery. "Christian ethics," the Lutheran theologians said, "cannot make a categorical position disapproving of artificial insemination by donor. Nevertheless there are psychological, social, and legal reasons which might lead Christians to refuse artificial insemination by donor for themselves. Finally, of course, the decisions rest with the persons who are involved."

Other statements by mainline Protestant denominations run about the same. I know of no fundamentalist church body which has spoken on the matter. When questioned on the subject, Billy Graham responded, "The procreative process was devised by an infinite, wise God, and is . . . sacred. The Scriptures mention no other method of reproduction of the human species. . . . When the Scriptures are silent, there is a show of doubt."[6]

In England, the Anglican Archbishop of Canterbury appointed a special commission of theologians to study AID. The result was this statement:

> The strongest plea put forward in defence of AID is the plight of the married women who long for children but whose husbands are sterile. . . . We need not affirm once more the profound compassion which this frustration must evoke; but our concern at this point is to answer the question whether such a desire . . . [is] in strict truth *inordinate:* one which exceeds the proper bounds of desire . . . We are all without distinction, required to restrain our desires, however imperious. On what rational ground is it urged that while *sexual* desires ought not to be indulged at all, parental desires may be?[7]

Helmut Thielicke, a European Protestant theologian, is likewise opposed to AID. He holds with Catholic moralists that marriage conveys exclusive bodily rights to sexual and reproductive organs and that AID would be a violation of these rights. He also claims that AID threatens marriage because it fulfills motherhood at the expense of confirming failure of the father. Fidelity in marriage, he insists, is first a personal bond between husband and wife and not primarily a legal contract. Parenthood is "a moral relationship with children, not a material or merely physical relationship."[8]

As we might expect, Joseph Fletcher sees no moral problem with AID. For one who denies the authority of Scripture, he takes the curious stance that AID is permissible because polygamy, concubinage, and levirate marriage occurred among God's Old Testament people. He notes that Leah and Rachel sent their husband Jacob to conceive children by their handmaids (Gen. 30:3-10), barren Sarah dispatched her husband Abraham to her fertile maid Hagar (Gen. 16:2), and Onan displeased the Lord by refusing to impregnate Tamar, the wife of his dead and childless brother Er. (Gen. 38:8-10).

There are other reasons, besides adultery, that make AID questionable. One is the deceit involved in naming the husband as the father on the birth certificate. I'm old-fashioned enough to believe that a lie is a lie. Furthermore, it's wrong from my viewpoint to deceive an obstetrician into thinking that a woman is pregnant by her husband so that he will declare the husband the father.

To a Christian couple contemplating AID, I would say, "Have you considered that it may be God's will for you not to have children of your own? Might he not have another child for you to love, or some other outlet for the care and time which you would give a natural child?"

I know that it isn't easy to find a child for adoption. Tens of thousands of babies who might have been adopted have instead been aborted. Most unwed mothers who choose not to abort are keeping their babies because the public stigma is not as great as it once was. Still, I think a couple ought to look into adoption. It might be expensive, but then look at what it costs to have a child in a hospital. If adoption doesn't work out, there is the possibility of caring for a foster child or of sponsoring a child overseas through a missionary agency.

The late Dr. H. J. Muller, a Nobel Prize winner in physiology and medicine, had a plan to carry the concept of AID a step further. He called on prospective parents to forego "egotistical" desires to reproduce their own heredities, and instead, to help advance the human race by constructing children from the "best" available egg and sperm.[9]

This is the same old song of humanistic evolution. With the development of IVF and the prospect of cloning added to AID, we're now hearing more about Dr. Muller's scheme.

The Muller plan is to store the best sperm and eggs in banks until the political climate permits establishment of experimental farms. A California sperm bank containing the seed of several Nobel prize winners was recently named for him. His widow has protested the use of the Muller name.

Frightening? You bet. But nontraditional methods of reproduction through cloning, IVF, and AID are only one part of the picture. In the next chapter we'll start delving into the world of genetics and look at procedures which may be used to change human heredity itself.

THE
HEREDITY
TRAP

A child bounces happily on his father's lap. The mother looks on fondly at her two look-alikes. The freckles on the bridge of their noses are the same. They have the same jutting chin, even the same quick laugh.

Credit heredity.

A less familiar scene is not so inspiring. A two-year-old sightless child screams in pain. His parents try to comfort him while the nurse is preparing a narcotic. Their faces are strained, reflecting months of mental anguish. Their child is a victim of Tay-Sachs disease, which afflicts an average of one in 3,600 Jewish males. Merciful death usually comes before age three.

Mark this down to heredity also.

Huntington's chorea is also inherited. Because it appears later in life, many victims already have children by the time the disease appears. The children must then live with the agonizing uncertainty of whether or not they, too, will have the disease. About 25,000 Americans suffer with this affliction.

More is known about Huntington's disease than about other genetic diseases, perhaps because it killed folksinger Woody Guthrie. Victims lose control of facial and body movements, one muscle at a time. They frequently lose their balance and fall. Their arms begin flailing at the most embarrassing moments. Severe emotional problems, depression, and violent behavior may result. Many victims commit suicide.

Muscular dystrophy, marked by progressive weakening and degeneration of muscle fibers, affects about 200,000 children and youth in the U.S. It is actually a class of diseases, the most common being Duchenne dystrophy which is thought to be inherited by boys from their mothers, who themselves do not have the disease. The victim appears normal at birth, but before six he is waddling, walking on his toes, and falling frequently. By twelve he will probably be in a wheelchair. By twenty he will likely be dead from a respiratory infection that his breathing muscles were too weak to combat. Around 60 million dollars is spent each year on physical therapy, drugs, and research related to Duchenne and thirty-seven other neuromuscular diseases. No miracle cure has been found. At best, only certain symptoms can be controlled and victims helped to live as normal a life as possible.

Much is yet to be learned about the origins of the various dystrophies. The consensus among researchers is that heredity is probably at fault.

Here's a young man of twenty-eight, father of two, and just starting his career. He loves sports and is the picture of health. One day he keels over while playing tennis. His partner speeds him to a hospital where the doctor pronounces him dead of coronary thrombosis. A coronary artery had become hard and narrow because of arteriosclerosis. A blood clot had blocked the artery, cutting off the blood supply to part of the heart.

Arteriosclerosis—caused by the buildup of fatty deposits and calcium—is not unusual in older men. Why is it present in such a large degree in one so young? Doctors speculate that the young father suffered from hyper-cholesterolemia, a genetic disease which involves a deficiency in the chemical agents that normally hold down cholesterol. Left unregulated, the cholesterol caused fatty deposits to build up rapidly in the arteries.

Each of us begins life as a combination of the heredity of our parents, which they in turn inherited from their parents. Most of us are born relatively healthy. I say relatively because no one escapes a hereditary handicap in some form. It may be as harmless as flat feet or as devastating as SLE (systemic lupus erythematosus—a connective tissue disorder of internal parts of the body); a predisposition to a more familiar disease such as arteriosclerosis, or emphysema; or a combination of factors which will never be diagnosed by medical science. I'll say it again. No one is born in *perfect* genetic health. Each of us receives some bad heredity.

When you hear the term "inherited disease," you might assume the problem is too complicated to understand. However, the basic principles of genetics are not all that hard to grasp. The exceptions and the combinations of factors involved may be baffling, but this shouldn't keep you from understanding how the laws of heredity ordinarily work.

Is it really essential to know about chromosomes, genes, and DNA when genetic disease hasn't touched you or one of your loved ones? It is if you want to understand what today's scientists have in mind for the reshaping of the human race. If you're interested in the heredity of your posterity and why many scientists want to change it, you'd better care. You may decide that the new genetic engineers are merciful. Or you may conclude that their research is diabolical and should be banned altogether.

Class is again in session. This time, we'll start with a history lesson. The father of modern genetics was a pea-growing Roman Catholic monk, named Gregor Mendel, who lived in the last century. Before his discoveries were made known, people could only speculate about how heredity was passed on.

Aristotle, for example, believed sperm contained all the plans for an offspring. He was at least half right. Mendel proved, among other things, that the characteristics of offspring are shaped by both parents and follow a predictable pattern, generation after generation.

Mendel discovered that crossing yellow-seeded peas with green-seeded ones always produced a totally yellow first generation. But the green color always returned in the second generation, in one of every four offspring. This led him to two conclusions: (1) certain characteristics are dominant over others when different parents are crossed, yellow over green, for example, with the characteristic of only one of the parents showing up in the offspring; (2) other characteristics are recessive, or secondary, but they are not wiped out in the crossbreeding. They reappear in succeeding generations. In one experiment after another, Mendel confirmed that this kept happening. He showed that the reappearance of characteristics, or hereditary traits, could be predicted mathematically, generation after generation.

Mendel's findings were printed in a scientific journal in 1866, but were ignored by scientists—perhaps because the monk's discoveries contradicted the theories of another scientist named Charles Darwin. Darwin believed that acquired characteristics could be inherited and that change was a normal process. By Darwin's reasoning, the giraffe's long neck was "acquired" from stretching to reach leaves on high trees, which evolved, in time, into a hereditary pattern.

While Mendel's discoveries gathered dust, western scientists were coming to believe that heredity was transmitted through the sperm and egg that fuse to produce the first cell of the new life. By conducting experiments with fruit flies, Thomas Morgan unknowingly confirmed many of Mendel's conclusions, while demonstrating that exceptions sometimes did occur in the mathematics of transmission.

Mendel's report was finally "discovered" about 1900 The precepts which the monk had so carefully set down demonstrated what many western scientists now suspected. He became a scientific hero in the United States.

It was a different story in the Soviet Union. A botanist named Lysenko convinced Stalin and his Communist party bosses that Mendel's laws were all a part of an imperialist plot. Acquired traits, said Lysenko and his friends, were transmitted to offspring. If you cut the tails off dogs for enough generations, they insisted, a dog would eventually be born without a tail. The ideas of the "pea-picker" (Mendel) and the "fly-breeder" (Morgan) were officially condemned by the Central Committee of the Communist Party. Because of such dogmatism, the Russians would remain far behind the west in knowledge of genetics. This has vastly affected their method of agriculture and partially contributed to their continued poor grain harvests. They still have not caught up with the west in genetics.

Our knowledge of human genetics keeps advancing rapidly. Scientists can now identify about 2,500 genetic maladies. This includes such rare inherited diseases as hemophilia and cystic fibrosis and some forms of more common diseases such as cancer. Researchers at the Creighton School of Medicine in Omaha, Nebraska, have just announced findings that indicate between 10 percent and 20 percent of all varieties of cancer are transmitted from generation to generation. Breast cancer heads the list.

Genetic diseases now known can roughly be placed in three categories:

First are those diseases caused by abnormal numbers or shapes of chromosomes. Chromosomes, you'll remember, are found in the cell nucleus, twenty-three pairs to a set, with each parent contributing one in a pair.

An example of this type is the familiar Down's syndrome or mongolism. This defect usually results from

97

having an extra twenty-first chromosome, three instead of the usual two. This is more likely to occur with the child of an older mother. We don't yet know why, although recent research relates it to the decrease of female hormones. Other research has shown that the risk is also increased for older fathers.

Another defect from chromosomal abnormality is Turner's syndrome, which prevents a child from developing secondary sex characteristics at the age of maturity. This disease is marked by the presence of a single X or twenty-third chromosome, with no X or Y partner to produce a normal female or male. Sometimes a problem may be caused from having three X chromosomes, when there should have been two, although the abnormalities in this condition are so minor that they rarely come to a physician's attention.

The second group of genetic diseases is caused by mutations, chemical mistakes in the construction of individual genes. (Genes, as we've already said, are contained in chromosomes and are matched in pairs accordingly.) Many diseases fall into this category. Hemophilia and Duchenne dystrophy are two which I've already mentioned. Another is PKU (phenylketonuria), a metabolic disorder which causes mental retardation unless treatment begins shortly after birth. Another is galactosemia, which prevents a baby from metabolizing the milk sugar, galactose. If given milk, the child may die within two weeks. A change of diet will prevent tragedy.

The third class of inherited disease are those which are influenced by a number of genes and may be set off or pushed forward by environmental factors. Hyper-cholesterolemia, for example, is aggravated by a diet rich in cholesterol. A low-cholesterol diet may prevent or at least postpone a coronary.

Researchers at a Veteran's hospital found that 50 percent of their emphysema patients had an inborn shortage of the biochemical alpha-1-antitrypsin. The chemical ordinarily fights pollution and foreign substances in the lungs. A person with this deficiency who protects his lungs against cigarette smoke and other dangerous pollutants may never suffer emphysema. Thirteen times as many smokers as nonsmokers develop emphysema. Their lungs become less elastic until finally they are literally unable to breathe.

We're discovering more and more links between certain genetic diseases and environment. Application of knowledge in this field could increase life expectancy several more years, yet some scientists are not pleased that modern medicine is learning how to control genetic disease. Here's why. There was a time when persons with genetic diseases didn't live long enough to bear children. Consequently, they didn't pass on their problem heredity. Today, they're living longer and raising families. Their bad heredity is multiplying and spreading, resulting in an increase of genetic defects in the population.

This problem has also been aggravated as medical science has wiped out or effectively curtailed one killer disease after another. More soldiers, for example, died from typhoid fever during the Civil War than were killed in battle. When did you last hear of someone dying from typhoid? The flu epidemic of 1918-19 took more lives than World War I. In 1978 only about 7,000 influenza deaths were recorded. Most of these were elderly people.

Typhoid and influenza are not genetic diseases. But these and other non-genetic maladies tend to take a greater toll among persons with hereditary deficiencies.

To illustrate further: Once persons born with poor or failing eyesight had a hard time surviving. Some got picked off by wild animals. Some may have been run over by chariots. Now we have eyeglasses, contact lenses, and remedial surgery. Most people with bad eyes live and enjoy life as long as those with good eyes, and keep increasing the bad genes in the population. This is good for optometry but bad for genetics.

A cold-blooded scientist will argue for a population control that calls for killing off the genetic weaklings or sterilizing them so they can't bear children. This is the only way, he says, to have a healthy race.

He's probably right.

We know this has happened in some societies. One that comes immediately to mind is the Aucas, the Indian group that killed five American missionaries trying to contact them about twenty-five years ago. The Aucas were then probably the most isolated tribe in South America. No outsider had ever lived among them.

Much has happened since the five missionaries were killed. Elisabeth Elliot, the wife of one of the martyrs, and Rachel Saint, the sister of another, entered Auca country with a young Auca woman who had come to the outside. The very men who killed the missionaries were converted. Most of the tribe are now Christians.

Before Christianity came, it was the custom of this small tribe to kill weak babies, usually by choking them to death or burying them alive. The missionaries taught them to spare these little ones.

The "weak" children that have been saved are just now getting old enough to have children of their own. There has been no noticeable genetic change as yet, making it possible for medical researchers to determine the well-being of a group that had habitually killed off weaklings.

A team of doctors from Duke University Medical School visited the Aucas and discovered no evidence of hypertension, heart disease, diabetes, or cancer. The highest blood pressure they could find was a systolic reading of 109 millimeters. (A reading of 120 is considered normal in our population for a twenty-year-old male.) This person, an older man, was upset about something when his pressure was taken.

The Aucas will probably not remain this healthy for long. The weak ones will have children, thus spreading bad heredity in the group. The tribe can be exposed to civilized diseases. This has already happened. In 1969 a polio epidemic swept through the tribe, killing sixteen. A Quichua Indian from the outside was blamed for bringing the disease. The Aucas have since been vaccinated by the missionaries, who are extremely careful about exposing any of the Indians to infectious maladies.

Looking at the Auca experience from a scientific point of view, does it make sense for the missionaries to be there? Many non-Christians will say no. But there are a couple of other things I haven't mentioned. First, the Aucas and other "primitive" tribes are going to be exposed to civilization, whether missionaries go to them or not. It's just a question of who is going to help them face the modern world.

Second, it has been determined that before the missionaries came, the Aucas did have a high death rate. Over 50 percent of this mortality resulted from spearings in revenge raids. Since the missionaries came, these raids have stopped and the population has doubled.

So, at least with the Aucas, there are more factors to consider than genetics.

The pessimistic scientists, however, are looking at the race's genetic future from a wider perspective. They say that the way mankind is going, saving lives and allowing genetic defectives to live and reproduce, we're headed toward a genetic Armageddon.

They predict this from simple laws of probability. With more people alive carrying bad genes, there is more likelihood of disastrous combinations in marriage, resulting in more defective children. Let's simplify this a bit: A hundred years ago ten persons in a thousand might have carried bad genes for a certain genetic disease which would not appear until two carriers married. The chances of this happening were pretty small. Now the ratio of carriers to population could be fifty in a thousand, increasing the probabilities greatly.

Let's get down to some basic facts about how bad genes

can interact in a marriage to produce a child with genetic disease. Genes, as we've said, come in pairs. In each pair one gene is dominant and the other recessive. Let N represent a dominant healthy gene; n an abnormal recessive gene. A pair of two normal genes would be written NN. A person with such a pair is called a *homozygote* (*homo* means same; *zygote* is the cell that results from the fusion of sperm and egg). A normal and an abnormal gene in a pair would be written Nn. A bearer of this combination is called a *heterozygote* (*hetero* means different).

The marriage of two normal homozygotes (NN plus NN) almost always produces another NN in their offspring. Only in rare instances will outside influences (such as radiation) damage the character of a gene, changing a good N into a bad n.

The mating of two heterozygotes can spell trouble. For example, in the Mendelian formula for peas, when dominant yellow (N) and recessive green (n) are paired in both parent pea plants there is one chance in four of producing a green (nn) offspring. Peas and people act alike. If a man with Nn marries an Nn woman, each child will have one chance in four of receiving two ns. With no good N to control the bad n, a defect will result at this spot in the gene chain.

Let's apply this to cystic fibrosis, a disease which causes malnutrition and an excess of mucus secretions throughout the body. Its victims often die from lung infections. Presently one in every twenty-five persons in our population is a carrier of one of the genes which causes cystic fibrosis. This person is Nn, the little n standing for the bad gene. If the carrier marries a normal person, the formula for each of their chidren will be NN or Nn and none of their children will have the disease. The dominant N will checkmate the recessive n. But suppose a carrier marries a carrier ($Nn \times Nn$). In this case one-fourth of their children (on the average) will be born with cystic fibrosis.

Or consider hemophilia, one of the so-called sex-linked

102

diseases. It's called that because it is the result of a gene carried on the X chromosome.

The cells of a female each have two X chromosomes, while those of a male have one X and one Y chromosome. A female with a bad sex-linked gene will have one X that is n and a healthy X with N. The healthy N would over-balance the n. She will only be a carrier. But the male has only one X chromosome, so if he inherits the bad n there will be nothing to counteract it and he will have the disease. His Y chromosome does not contain the same gene.

All of this may seem like an alphabet game. But it is agonizingly serious to families who have children suffering from genetic diseases. It happens in tens of thousands of families every year—rich and poor, mighty and lowly, just and unjust; genetic disease plays no favorites and operates according to the laws of heredity, which are rarely fickle. And, as we've already said, it's getting worse because bad genes are increasing faster than the population as carriers live longer and pass on their potentially lethal heredity.

Evolutionists are concerned by this, but they're not surprised. They believe that if we keep the genetically defective alive to reproduce, genetic diseases will keep increasing until the species called *Homo sapiens* is wiped out. Even if we were to allow the diseases to take their toll, they feel, mutations will accumulate and turn us into something very different. Dr. James Bonner, a biologist at the California Institute of Technology, put it this way: "The normal expectation of an animal species such as our own is to arise through mutation, evolution, and selection, and then to die out and be replaced by a species more fitted to the then current environment." Bonner believes that mutations—genetic mistakes in the transmission of heredity—have already formed millions of species of different organisms since life boiled up in the sea over three billion years ago. Most of these species, he thinks, have been wiped out and replaced by hardier organisms.[1]

Optimists among the evolutionists hope that man can

rise above the past, grab the reins of the evolutionary force which is dragging him down, and create a new humanity through science and understanding. Pessimists think this is not likely to happen. Some suggest gloomily that there may be no *Homo sapiens* a few centuries from now.

What do the hopeful evolutionists plan to do? From their scientific viewpoint, the sensible thing would be to identify, then abort or kill off all weak babies at birth, while forcing known adult carriers to be sterilized.

This couldn't be done today—at least not in our society. I'd like to think that scientists wouldn't do it even if they could. I believe they've been influenced more by Christian values than they will admit. In the first century, Greek and Roman pagans threw unwanted babies on garbage heaps. Christians picked them up and started the first orphanages. Christians are still caring for the unwanted today.

The evolutionists have another game plan—genetic engineering. This is to be carried out on two fronts. On one hand, they anticipate an increase in abortions as parents-to-be become aware of defective embryos through prenatal genetic tests. On the other, they hope that new research on gene transplants and artificial DNA will make it possible to change bad heredity into good. They expect the new genetic engineers "to take over Mother Nature's job of genetic card shuffling, with clear intentions of stacking the deck in man's favor."[2]

How are we to respond as Christians? Should we oppose prenatal genetic testing? Should we be against gene transplants on the basis that this is tampering with God's heredity? Can we approve of some forms of genetic engineering, while disapproving of the evolutionary goal of building a super species?

These are questions which we must face.

9

SCIENTIFIC "BREEDING"

Ten-year-old Monica Anderson's body appears to be turning to bone. A hard crust has grown over her muscles and joints. She is in constant pain and every breath is a struggle because her rib cage is hardening over her lungs. Dr. Michael Zasloff, a geneticist at the National Institute of Health in Bethesda, Maryland, where Monica is hospitalized, says she might survive only a few more months or she may live to be sixty. The disease (classified as *myositis ossificans progessiva),* he adds, is so mysterious that it could stop as quickly as it started.

Across the country in San Diego, a little girl just five years old recently died of "old age." Doctors at Children's Hospital there say Penny Vantine aged at a rate of fifteen to twenty years every twelve months. At the time of her death, her hair was dry and sparse, her face drawn. The skin was nearly transparent, showing the veins in her forehead. She had glaucoma and cataracts in both eyes. Her circulation was poor, her blood pressure was high, and her fingers were swollen with arthritis. The doctors know no cure for this disease which they think came as a result of a disorder in the child's metabolic or endocrine system, probably caused by a genetic defect.

It is the verdict of "no cure" that makes genetic diseases so heart-rending. Medical science can only treat the symptoms and help victims live as comfortably as possible. In some instances, control therapies can help persons live almost normal lives, but often little can be done to slow the ravages of mind and body that result from inherited afflictions.

Until recently most research money has gone for developing new drugs and other therapies that can control genetic illnesses. Now, thanks to new knowledge of genetics, more emphasis is being put on preventing the birth of children with genetic defects.

More people are also being trained in genetic counseling. This service has been available at a few university hospitals for several years, but was little used because of lack of public awareness. Now, I expect every gynecologist is recommending it to patients with a high risk for bearing defective children.

Spinal bifida is a genetically caused birth defect in which the end of the spinal cord has little or no covering. It causes varying degrees of paralysis of the legs, bowels, and bladder. In most instances, excess spinal fluid also collects in the head, causing hydrocephalus which, if not treated, results in mental retardation. Each year, twelve thousand children are born with this defect in the United States.

The counselor can say that the general risk in the population is 1/600. But the risk for recurrence following the birth of a child with an open spine is a foreboding 1/25. Having a previous child with a genetic defect is a frequent reason for seeing a genetic counselor.

A more complicated case is two first cousins who want to marry in a state where such marriages are permissible. The man has a retarded brother with phenylketonuria (PKU). He and his fiancée consult a genetic counselor about the possibility that they might have a child with PKU. The counselor says, "You are wise to be concerned.

106

Couples who are related run a much higher risk than those who are unrelated." Then he writes on his chalkboard:

Frequency of PKU carriers in general population = 1/100
Chance of two carriers marrying = 1/100 × 1/100 = 1/10,000
Chance of two carriers having PKU child = 1/4
Chance of unrelated couple having PKU child = 1/10,000 × 1/4 = 1/40,000

Directing his attention to the man, the counselor says, "Because you have a brother with PKU, you run a higher risk no matter whom you marry." And then he writes:

Chance of you being a carrier = 2/3
Chance of you marrying a carrier = 1/100
Chance of PKU child = 2/3 × 1/100 × 1/4 = 1/600

"Now in the case of you and your fiancée," says the counselor, "the risk is even greater." Again he turns to write:

Chance of you being a carrier = 2/3
Chance of cousin being a carrier = 1/4
Chance of PKU child = 2/3 × 1/4 × 1/4 = 1/24

Until recently, about all a counselor could do was take a family history and figure probabilities on the basis of Mendel's laws and other related knowledge. Blood tests can now identify the carriers of some diseases and prenatal tests can tell if an unborn child has certain genetic defects. The prenatal tests are by far the most controversial, since findings frequently influence a couple to have an abortion.

The most widely administered prenatal test is amniocentesis. It is usually done between the fourteenth and sixteenth week of pregnancy. A long needle is inserted into the amniotic sac, where the baby is developing, and a sample of amniotic fluid withdrawn. (There is a 1 to 5 percent risk, depending on the skill of the surgeon, that the needle will injure the fetus or stimulate a spontaneous abortion.) Fetal cells are then taken from the fluid and grown in a tissue culture for seven to twelve weeks. By the time the cells have been examined, the pregnancy is twenty-one to twenty-eight weeks advanced.

Amniocentesis, for example, can enable the genetic counselor to tell a couple, with a less than 1 percent error margin, whether or not their baby has Down's syndrome (mongolism). Without the test they could only be told their probabilities of having a Down's syndrome child: 1/1500 for a twenty-five-year-old mother, 1/300 for a thirty-five-year-old mother, 1/25 for a forty-five-year-old. Many gynecologists advise pregnant women over thirty-five to have the test for this reason. In some states it is offered free to any pregnant woman past thirty-five.

In another situation a couple is concerned about hemophilia. The bad gene for this disease is passed from a carrier mother to her child on a 1/2 risk ratio. However, only a son can get the disease, while a daughter can be a carrier.

In this case the couple has come to the counselor because the pregnant wife has learned that her only brother died at age seven from uncontrolled bleeding. Until reading a magazine article, she had never thought there was any familial connection.

Amniocentesis will tell this couple if their unborn child is a boy. If so, a further blood test will indicate if their son has hemophilia. The blood will be checked for a critical clotting substance called Factor VIII. If this element is not present in sufficient amounts, the child's blood will be unable to clot and he may die of bleeding from the slightest wound.

Before amniocentesis became available, parents could

not find out if their unborn child suffered from Down's syndrome, hemophilia, or from about a hundred other genetic diseases which can now be identified through prenatal testing. With infanticide then being out of the question, they had to accept the child's affliction. Some parents kept mongoloid children. Some had them institutionalized at private or public expense. Hemophiliacs had to be watched constantly or given injections of synthetic Factor VIII at tremendous cost.

With amniocentesis, parents can know the child's state of health before birth. If the test results for some genetic disease are positive, the possibility of abortion is held out before them.

There is more to consider than the interests of the parents and child. Taxpayers may have a stake in their decision. The cost of caring for children with severe genetic defects in public institutions is staggering. The time may come when pregnant women will be required to have amniocentesis and if the test results are positive, to have an abortion.

Insurance companies are also very interested in genetic testing. Some companies may decide to insert clauses in future policies protecting themselves against claims, unless policyholders agree to prenatal testing and abortion if test results prove unfavorable.

Geneticists tend to be overwhelmingly in favor of such tests. Says Dr. Sara C. Finley, co-director of the Laboratory for Medical Genetics at the University of Alabama, Birmingham: "Many women with histories of genetic disease would be afraid to have a baby without amniocentesis."[1]

Those who oppose abortion tend to oppose prenatal genetic testing. Perhaps the most outspoken physician is Dr. C. Everett Koop, chief surgeon of Children's Hospital in Philadelphia. Dr. Koop is world-famous for his surgical skills in saving the lives of newborns with congenital malformations. He recently teamed with Dr. Francis Schaeffer to make the book-film series, "Whatever Happened to the Human Race?"

Dr. Koop claims the whole "system" of prenatal testing "is to find out if there is something wrong with the fetus. And if the fetus is defective some parents will decide to abort it. Since I take a high view of life I see amniocentesis as a search and destroy mission."[2]

Many supporters of amniocentesis are troubled by couples who use the test to determine the sex of a baby and then abort a child not of their choosing. Some doctors refuse the test to couples who admit this is their purpose. Dr. John C. Fletcher of the National Institute of Health believes doctors cannot prevent patients from having the test for this reason. "If you take the position on abortion that the Supreme Court takes, you can't logically and consistently uphold refusing these people."[3] Dr. Fletcher was co-chairman of a survey released by the Hastings Center in January, 1979, that reviewed the uses and the availability of amniocentesis. The Center estimated that 15,000 women took the test in 1978. This number is expected to increase rapidly.

The battle over abortion has centered around a question of whose rights must prevail, the right of the mother over her body and the new life she carries or the right of the child to live. Those who oppose abortion on demand believe that fetal life is sacred and should take precedence over any inconvenience, short of death or serious injury, to the mother or anyone else. Pro-lifers appear to be gaining ground. Recently one of the nation's leading pro-abortionists, Dr. Bernard N. Nathanson, radically changed his position. He is now calling for the outlawing of abortion or for a constitutional amendment protecting life in the womb except when the mother's life is in danger.

Yet many biological scientists think that abortion of identifiable genetic defectives will improve the human species. Dr. Cecil B. Jacobson, chief of the Reproductive

Genetics Unit at the George Washington University Hospital in Washington, D.C., is one of the early developers of amniocentesis. He clearly sees it as a tool for fashioning a better race in the future. He is for aborting mongoloids 100 percent, or any other child which parents don't want, including a healthy child whose parents prefer the opposite sex. "I just don't recognize any absolutes here," he says. If prenatal testing can be developed to show genetic pre-conditioning for diseases such as cancer, he would abort these children. "If we could tell what fetuses are going to be afflicted with cancer in their forties or fifties, I would be for aborting them now. That would eliminate some types of cancer forever."[4]

Does that shock you? Well, many believe we can only have a perfect society, a "brave new world," by getting rid of the unfit before they are born. Who will decide the criteria for fitness? Parents are supposed to make those decisions now. Yet recent studies indicate that they are greatly influenced by doctors who do the testing. Dr. John Fletcher followed twenty-five couples of varied social, ethnic, and religious backgrounds who were told their children were defective. After "counseling" all twenty-five opted for abortion, then were sterilized.

Some genetic defects now identifiable can be modified or controlled before birth to prevent further deterioration in health. Certain enzyme deficiencies can be detected in followup to amniocentesis. One of these deficiencies, methylmalonic acidemia (MMA), is marked by a lack of vitamin B12. If not treated, the child will be mentally and physically retarded at birth. Happily, before the birth of an MMA child, the mother can be given daily injections of B12 which will be absorbed by the fetus. After birth, the child can be put on a low-protein diet and given daily doses of B12.

Postnatal medical advances are making more control therapies possible for defectives. PKU, for example, can be detected at birth from a pin prick of blood or a drop of urine. The disease is caused by inadequate metabolism of protein. A special diet will usually check the retardation which may otherwise result. Testing for PKU is now required in many states and strongly recommended in others.

Some detectable genetic diseases may require surgery if death is to be averted. One is familial polyposis of the colon. Multiple small benign polyps begin showing up in the colon, although no outward symptoms may occur. This portends cancer of the colon by age forty or fifty. Half of the siblings and children of victims will develop the polyps and eventually have the cancer. The only way to prevent the cancer is to remove all or part of the colon in a colectomy when the polyps are still benign—drastic treatment for a symptomless patient.

Dr. Hymie Gordon, a consultant in the Medical Genetics Section of the Mayo Clinic, discovered this disease in a large South African family. He located four hundred "first-degree" relatives (children and siblings) of affected persons, half of whom were due to develop the polyps and subsequent cancer. At his recommendation, members of the extended family began coming in for examination and possible colectomies.

Such treatments do not cheer the scientists bent on "saving" the race from genetic doom. By saving the lives of victims, they say, we are further polluting the population gene pool. If left alone, many genetic defectives might not live to bear children. At the very least, the genetic engineers argue, the worst genetic defectives should be singled out and sterilized.

This gets into eugenics, the science of race improvement. Eugenics (literally "good genes") calls for selective mating to produce well-born children. Southern slave owners before the Civil War sought to improve their slaves by having choice young black men impregnate female slaves of childbearing age. Hitler bred selected young German Aryan men and women to produce superior children. Modern agriculturists and cattle raisers practice precision eugenics in improving their crops and herds.

Francis Galton (1822-1911), an Englishman, is called the "father of modern eugenics." A cousin of Charles Darwin, he thought that club feet, curvature of the spine, and a high-arched palate were inherited marks of criminality. Drunkenness, epilepsy, and poverty, he felt, were other evidences of "bad seed." He proposed laws to prevent persons displaying such bad characteristics from mating. He suggested that "healthy" young men and women be given certificates of merit and encouraged to reproduce. Christians were appalled by his ideas, but many English atheists and agnostics climbed on his bandwagon. Playwright George Bernard Shaw called Galton's proposal the only "religion that can save our civilization from the fate that has overtaken all previous civilizations."

Galton's idea fired imaginations in both England and the U.S. The number of mental patients sterilized in the U.S. will probably never be known. Most of the incriminating records were destroyed. In a single Kansas home for boys forty-four "unworthy" young men were castrated on a single occasion.

The Indiana Legislature passed a law in 1907 requiring sterilization of idiots, imbeciles, and the feeble-minded. Similar actions were carried out in other states, with and without legal sanction. In Virginia relatives of a young "feeble-minded" woman engaged lawyers to defend her rights. They appealed the case to the U.S. Supreme Court on grounds the state courts had denied her equal protection of the law under the Fourteenth Amendment. "Three generations of imbeciles are enough," declared Chief Justice Oliver Wendell Holmes, in supporting the Virginia courts. The young woman was sterilized.

The Nazi horror during World War II stirred public revulsion against human eugenics. But the idea for keeping "unfit" people from reproducing did not die. In 1965 a report on criminal patients at Carstairs Maximum Security Hospital in Scotland claimed that 3.5 percent of the male inmates had an extra Y chromosome. The Carstairs study was not taken seriously until a story surfaced that chromosome analysis had revealed Richard Speck, the Chicago mass murderer of nine nurses, to be an XYY. Coincidentally, researchers at Harvard and in Sweden developed special fluorescent stains that made it possible to read some chromosome patterns in cells taken from the inner lining of the cheek.

Voices were raised calling for mass screening of newborn infants to detect criminal tendencies. But before the scheme could be put into action, other scientific studies were published that contradicted the Carstairs results. One researcher reported finding XYYs among a group of businessmen, clergymen, and factory workers. The embarrassed promoters of screening conceded the error and backed away.

About the time the XYY scheme was put to rest, a clamor arose among blacks in the American civil rights movement for government funding to research sickle cell anemia. Support of such research, they said, would indicate how much America was committed to the cause of helping minorities. In the propaganda that followed, the distinction between sickle cell carriers and victims became blurred.

Eight percent of all blacks were said to be carriers of the trait that "threatens to cripple or kill" (a frequently heard phrase). Newspaper articles and TV specials highlighted the complications of a sickle cell attack: oxygen-starved red blood cells would clump together and clog blood vessels, this would result in painfully swollen joints, damage to kidneys, and lowered resistance to infection. Twenty-nine states and the District of Columbia quickly enacted voluntary screening programs. Some influential politi-

114

cians demanded that the screening be made compulsory for all blacks.

This crusade backfired. "Are you a carrier of sickle cell?" became a standard question put to blacks in job interviews. Airlines stopped hiring blacks for flying duties on grounds that lowering of oxygen could cause a sickle cell attack. Some life insurance companies raised rates for sickle cell carriers.

Scientific research finally cleared the air. Few carriers would actually ever get the disease. Only under conditions of extreme oxygen lack was an attack likely to occur. None of the black African athletes at the high-altitude Mexican Olympic Games in 1968 had experienced a sickle-cell "crisis." No documented reports could be found of a single sickle-cell carrier having an attack during air flight. Yet the myth continues today and some blacks suffer job discrimination because of false information.

Man's record in eugenics is not very good. Still, the race improvers have not given up. Armed with refined lab technology and factual studies of real genetic horrors, genetic engineers are pushing for an expansion of screening and restrictions on marriage for the worst risks. They point out that some laws for eugenics are already on the books: It is illegal for brothers and sisters to marry in all states and some jurisdictions require sterilization of the severely retarded.

Their first big success has come from passage of the national Genetic Disease Act by Congress, creating a special unit within HEW. This unit is expected to provide more financial aid for research, training of genetic counselors, and public education programs concerning genetic disease. Additionally, many states are stepping up genetic screening. All babies born in New York State, for example, must be checked for five inherited diseases.

The necessity for education in genetics can readily be seen. We need to be aware of genetic problems in family lines. We need to know how to seek genetic counseling and determine the risk factors in producing children. When we have the facts, we should make responsible decisions about having children.

This responsibility should go beyond ourselves. If I have friends who produce a child with cystic fibrosis, and they do not understand how the disease runs in a family, I have an obligation to warn them that there is a one in four chance the next child born will also have the malady. Doctors should certainly know enough about the laws of genetics to know when to refer patients for genetic counseling. Pastors should have some awareness for premarital counseling.

I believe God is sovereign and carries out his will among us. Yet I don't think it's right to ignore obvious disease symptoms or flout common rules of health, and then charge the tragedy off to the will of God when the inevitable happens. The businessman who drives up his blood pressure to keep ahead of the pack is not, in my judgment, cooperating with the will of God. Nor is the young wife who knows she carries the gene for hemophilia doing God's will by getting pregnant and taking the grave risk of having a hemophiliac son or a daughter who is a carrier.

I personally recommend getting genetic counseling whenever there is any doubt. I am opposed to amniocentesis when intentions are to abort a child if the presence of a genetic defect is indicated. I don't think either hemophilia or Down's syndrome, which can be revealed by amniocentesis, offer grounds for abortion—certainly not for Christian parents. In any situation, parents should reflect soberly on the findings. If testing indicates that the child will be born seriously deformed, they should probably be prepared to expect a long period of hospitalization.

A genetics counselor is sometimes put in a predicament over how far to go in informing relatives of a patient. Take tylosis, a rare inherited skin disease easily detected by a mild rash of the palms of the hands. Seventy-five percent, on the average, of the people who have tylosis will eventually develop cancer of the esophagus which is almost always fatal. The counselor will give his patient this information. Should he depend on the patient to advise relatives that they should be checked for tylosis? Or should he notify them himself?

What is the counselor's responsibility when he has a patient with Huntington's chorea—a disease which doesn't become apparent until middle age? Again, should he depend on the patient to inform his family or pass on this information himself? Should the siblings be told at all, leaving them to live in suspended fear for years to come?

The counselor's job is to inform and not to make recommendations. But patients are going to ask, "What do you think I (we) should do?" If the patients are Christians, a Christian counselor can tell them to pray about their problem and perhaps talk to a trusted spiritual counselor such as their pastor. That may not be enough, even for Christians. Should the counselor also get into value judgments about whether a carrier of hemophilia should have children or whether two carriers of the cystic fibrosis gene should marry and reproduce?

Does the genetics counselor play God by helping the individual or couple reach a responsible decision about having a child? Do parents play God when they decide?

At this time almost all genetic decisions are a matter of individual choice. As better testing techniques become available, making it possible to identify more defects in the unborn and in carriers, more people will be faced with more choices.

The way our nation is moving at present, I foresee that some of these choices are soon going to be made by government decree. I fear that unless there is a dramatic turnaround, "Big Brother" will be telling many genetic carriers whether they can marry, whom they can or cannot marry, and if they can have children. It may reach the point that a couple must get a license to have a baby.

Still, we have not seen the whole picture of genetic engineering. Test-tube babies, artificial insemination, cloning, genetic testing, and abortion of the unfit all pale in comparison with the latest genetic research. This will be the subject of the next chapter.

10
RESHAPING
HEREDITY

During the summer of 1939 Albert Einstein wrote to President Roosevelt, urging him to investigate the possibility of building an atomic bomb. Einstein, a German Jew who had fled Nazi persecution, was then working at Princeton. He feared Hitler's scientists might develop the bomb first and conquer the world.

Spurred by Einstein's letter, Roosevelt authorized a crash project. Two years later a team of scientists, working at the University of Chicago, produced the first man-made chain reaction of atomic energy and unlocked the awesome power of the atom. Upon realizing what they had accomplished, one of the scientists reportedly reflected, "God would not have wanted this."

Many people feel this way about the biorevolution. They think scientists may be displeasing, perhaps even defying God, by developing sperm banks, conceiving babies in a laboratory, experimenting with cloning, and devising sophisticated tests which can determine a baby's sex and genetic health before birth.

But what about recombinant DNA experiments through which scientists are striving to reshape heredity itself? This is the newest frontier in genetic engineering. The potential

119

for changing the pattern of human life is far more important than artificial insemination, test-tube babies, and cloning combined.

This story illustrates what the recombinant DNA researchers are about. A mountain man came upon a city camper trying to get a bucket of clear water from a muddy stream. The native observed the visitor's futile efforts for a moment, then suggested, "Stranger, if you'd go up and run that hawg out of the spring, the water would clear up."

The more immediate goal of this new research is to clean up the "spring" of life by replacing bad heredity with good heredity, thus eliminating harmful mutations from the human gene pool, and perhaps even finding a way to stop the aging process. The long-range intent is to develop supermen with immunity to all diseases. "Such evolutionary developments," *Time* has suggested, "could well herald the birth of a new, more efficient, and perhaps even superior species. But would it be man?"[1]

The idea of reshaping man is being taken seriously by universities, government planners, and the media. One would have to be blind and deaf not to have heard something about DNA. Yet I find many people are puzzled, especially those who haven't had a biology course since the 1950s. One parent with a daughter in college studying microbiology put it this way: "This DNA code of life I hear her talking about makes about as much sense as the new math homework she brought home six years ago."

New math has been cut back in many schools. I can assure you that DNA is not going to disappear. If it did we'd all go with it.

Gregor Mendel, you'll recall, knew only that traits are passed from generation to generation with mathematical precision. By the time biologists began paying attention to his discovery, they suspected the hereditary information was inside the pairs of tiny threadlike strands in cell nuclei called chromosomes. They noticed that during cell division, the strands always split lengthwise, giving each off-spring cell a full share of the genetic material. Thomas Morgan confirmed this with his famous experiments with fruit flies.

By the 1940s biologists were assuming that heredity was in the pairs of genes strung inside the chromosomes. Back in 1871, a Swiss biochemist, Friedrich Miescher, had identified DNA (deoxyribonucleic acid) as being present in cell nuclei. But DNA's role in heredity was not understood until 1944 when scientists at the Rockefeller Institute mixed some DNA from genes of pneumonia bacteria with harmless bacteria in a lab dish. The harmless bacteria turned virulent, a demonstration that DNA carried a genetic message. But they couldn't understand how it had happened.

One day in 1953 two young scientists were overheard talking loudly in an English pub near their lab at Cambridge University. Someone asked James D. Watson and Francis Crick what they were so excited about. "We have discovered the secret of life," Crick shouted.

Crick exaggerated. But their achievement was still stupendous. Crick and Watson had built a Tinker-toy-like model of DNA based primarily on X-ray diffraction studies by other scientists, to explain the code by which the chemical directed the building of life. Their work won them the Nobel Prize and forced biology books to be rewritten.

The two researchers showed that DNA is shaped like a spiral staircase. The "banisters" are composed of long links of sugars and phosphates. The steps between them are made of four pairs of chemical bases, lightly connected at the center.

They coded the chemical rungs A-adenine, T-thymine, C-cytosine, and G-guanine. They showed that the sequence of A, T, C, and G could vary widely, providing an almost limitless information-storage system somewhat like the memory bank of a computer. Because A always pairs with T, and C with G, one side of the staircase was something of a genetic mirror image of the other.

In cell division, they depicted the DNA molecule as unwinding and unzipping down the middle of the staircase, with the pairs A-T and C-G breaking apart at the center. New bases replaced the missing halves of each strand, forming two identical copies of the original staircase. In this way DNA passed on its genetic orders to new cells and future generations.

Watson and Crick's explanation unleashed a deluge of new research projects. Biologists feverishly sought answers to more questions about the chemical nature of life.

The large protein molecules called enzymes were identified as the chemical engineers of life. Enzymes are long chains made from twenty different molecules, the amino acids. How did the DNA staircase order assembly of these acids into protein?

More studies revealed that sometimes, when DNA unwinds and unzips, instead of new DNA forming, a complementary messenger RNA (ribonucleic acid) forms to carry the genetic message. Messenger RNA goes to chemical interpreters (ribosomes) that "read" the building plans. As the code is read more coworkers, transfer RNA molecules, carry the appropriate amino acids to the ribosomes where they are hooked together into a protein chain.

How does DNA use a "telegraph" system of only four code letters (A, T, C, G) to select among twenty amino acids for producing complex proteins? George Gamow, a physicist, compared the four bases or rungs of the DNA ladder to different suits in a deck of playing cards. He proposed that DNA's four bases were "drawn" three at a time. Thus $4 \times 4 \times 4$ would yield sixty-four possible combinations.

Researchers at the U.S. National Institutes proved that three-letter DNA code messages could call up every one of the twenty amino acids, and even "punctuate" directions by marking the "end" and "beginning" of every sentence (job). This code is universal. The same four letters, used three at a time, specify the variety of protein-building amino acids in the nuclei of the cells of all living things (evidence that the same Architect drew all of the blueprints).

Armed with this new knowledge about the genetic code of life, scientists have since learned to transfer DNA from one species to another. They are recombining heredities at the level of design.

We've long heard jokes about breeding across the lines of species. For example: "What do you get when you cross an owl and a goat?" "A 'hoot-nanny!' "

Joking aside, this doesn't happen in normal reproduction. God so created each "kind" of creature that it is impossible for one kind to breed with another. Ten times in the first chapter of Genesis, the phrase "after their kind" is used. Usually only very similar members of a kind breed together. More rarely, quite different members of a kind may mate, such as lions with tigers or horses with donkeys—but always within the kind.

Recombination of DNA always occurs when a sperm fertilizes a egg. The new recombinant DNA research has started a brand new chapter by altering the order of creation. Some scientists who don't believe in divine creation are frankly scared. They're voicing the same fears that are surfacing about cloning: "Don't interfere with nature's basic way of transferring heredity."

Here is what has happened.

Scientists worked with *Escherichia coli,* a rod-shaped, one-celled microscopic bacteria. *E. coli* (ee-coh'-lye), as it is called, has long been one of biological science's most basic research tools. One reason is that it is so plentiful. It normally lives in the human intestine. Another is that, like other bacteria, it reproduces not by mating but simply by dividing. *E. coli* doubles about once every half hour. Having only one cell makes it a lot easier to study than an organism like man that has billions of cells.

Scientists found that *E. coli* consists of only a relatively few genes made of small closed loops of DNA, called plasmids. They discovered that when two bacteria touch each other, a connecting bridge sometimes forms and a plasmid from one passes into the other.

What would happen if plasmids from other bacteria were maneuvered near *E. coli*? Not only did the bacterium "invite" the visitors "in"; it also began reproducing, doubling, and redoubling, duplicating the recombined DNA each time. This meant that *E. coli* could serve as a miniature "pharmaceutical factory," developing "bugs" never before seen on earth. DNA could be mixed in the bacterium to form new combinations of bacteria.

Using *E. coli,* General Electric researchers developed a "bug" capable of breaking down a wide variety of hydrocarbons. They suggested a helicopter might drop a quantity on an oil slick and the bugs would multiply and lap up the oil.

During the excitement of achieving DNA recombination, few voices of caution were heard. Most of the talk was about the miraculous new substances recombinant DNA could produce cheaply. Insulin which is ordinarily taken from the pancreases of animals, was mentioned as one possibility. Another was Factor VIII, the expensive clotting material required by hemophiliacs.

Enthusiasm cooled when some scientists began to speculate that bugs formed from recombinant DNA might turn destructive. Someone wondered what would happen if the GE bacterium able to lap up oil slicks should get into an oil pipeline or the fuel tanks of a plane in flight. Someone else asked about the possibility of creating a killer bacterium in the lab which could escape and multiply millions of times before being missed. It might get into the water supply of a city like New York and infect masses of people. It might be resistant to every medical remedy known to man. It might wipe out humanity.

Scientists talked about such dread "what ifs," but nobody did anything until cancer researcher Robert Pollack heard that Paul Berg, a scientist at Stanford Medical Center was planning to insert a monkey virus, SV40, into *E. coli.* Pollack knew this virus had caused tumors when injected into laboratory animals and had made human cells cancerous in lab cultures, even though there was no record of it ever causing cancer in a human being. He called Berg, but the Stanford scientist saw no reason for worry. Berg felt his experiment was extremely significant. He reasoned that if science could understand why SV40 caused cancerous tumors, a big step might be taken toward understanding how cancer developed.

After talking with Pollack, Berg discussed the problem with his associates. They expressed concern that *E. coli,* armed with the tumor-causing virus, might get into somebody's intestines. Berg then agreed to stop his experiment.

Meanwhile a researcher in San Francisco, Herbert Boyer, found a better way to remove bits of DNA from cells. Back at Stanford, Stanley Cohen, a colleague of Berg's, discovered a versatile new plasmid which could easily pick up a strange gene and pass it off to *E. coli.* When this news hit the scientific grapevine, researchers from all over the world began calling and writing Cohen for samples.

Cohen told about his discovery at a meeting of 140 molecular biologists in New Hampshire during the summer of 1973. The scientists sat agape. Here was an easy way to splice any two kinds of DNA together. The possibility that someone would build a new life-destroying bug seemed much more fearsome.

An investigatory committee of scientists was appointed. They met at the Massachusetts Institute of Technology in April, 1974, and agreed to call for a temporary ban on recombinant DNA experiments considered dangerous.

This was followed by a think-tank conference of 134 scientists in California, plus a battery of lawyers and selected science writers. The scientists presented reports and discussed and argued the risks versus the potential benefits of recombinant DNA research. The lawyers warned that if a destructive bug should get loose and cause damage, the bug's developers could be sued. The scientists decided to continue the ban and to ask the National Institute of Health (NIH) to specify safety levels necessary for continuance of the experiments.

Stories written by the reporters at the conference about the possibilities of *Andromeda Strain* type bugs escaping and infecting the population created a huge stir. A special U.S. Senate committee met to hear evidence. Appearing before the senators, then Secretary of HEW Joseph Califano asked Congress to place federal restrictions on the exotic experiments. Alarmed, the NIH moved quickly to establish four levels of physical containment, ranging from P-1 for the lowest risk to P-4 for experiments with cells of higher animals and animal tumor viruses.

A few scientists continued to oppose all recombinant DNA research. Nobel Prize winner George Wald, Professor of Biology at Harvard, persuaded Mayor Alfred Velluci of Cambridge, Massachusetts (where Harvard and M.I.T. are located), to call a meeting of the city council on the subject. The council asked the two schools to halt all research being done at the P-3 level and above while a citizen's review board studied the situation. In February, 1977, the council decided that the experiments could continue, but only under safety standards more strict than the government required.

Fearing all recombinant DNA research might be stopped by an aroused public and government support funds cut off, scientists have been trying to develop ways to keep *E. coli* under control. Two researchers at the University of Alabama have developed a gene that makes it virtually impossible for *E. coli* to survive in a human body. But the worries continue and opinion is so sharply divided that research scientists at some universities are barely speaking to one another.

Many of the fears are justified. If a really dangerous bug is produced, there's a good chance it will escape. We have to reckon with the margin of error that seems to be part of every human project. McDonnell Douglas didn't build their DC-10 to crash. But apparently some of the workmen on the assembly line didn't do their job 100 percent. Even with that, the crashes that took over 600 lives might not have occurred. The manufacturer issued precise directions for servicing and maintaining the airplane. But the airline mechanics apparently didn't follow directions.

The scientists working on recombinant DNA assure us that all safety precautions have been taken. I'm still a little uneasy. I keep remembering that in 1974 sixty top nuclear scientists rated the chance of a serious nuclear accident about the same as the likelihood of a meteor hitting a major city—one in a million. On March 22 of the following year fire broke out at Brown's Ferry, Alabama, where the world's largest nuclear generating plant is located. Fortunately no one was hurt, but there could have been a major disaster with thousands of people killed. The plant was closed for three months. What caused it? A workman hunting for an air leak with a candle!

Since then we've had the scare at Three Mile Island in Pennsylvania. This accident could have resulted in a "melt-down," causing an explosion that would have ruptured the four-feet-thick concrete walls of the containment building and let out deadly radioactive gases. The dust around that accident has not settled yet and the defenders of nuclear power accuse the press of ballooning the story to get headlines.

Meanwhile, recombinant DNA research continues at a fast pace. Five large commercial companies, including Standard Oil of Indiana and National Distillers' Corporation, are heavily involved in seeking new disease-fighting remedies and new ways to grow food. DNA from fruits that resist cold weather, for example, might be transplanted into DNA of fruit that is subject to severe damage by frost. Oranges and grapefruit might be grown in Minnesota. "We're talking about billion-dollar possibilities," says Ronald Cape, chairman of the Cetus Corporation, the world's largest purely genetics firm.[2]

Research sponsored by another firm, Biogen, has succeeded in transferring the gene for human interferon into bacteria which now readily produce the protein. Interferon fights viruses and some types of cancer. Until now, it has been extracted from white blood cells through a process that produced only enough to treat 600 cancer patients a year at a cost of $10,000 each. Refinements of the new process may reduce costs to as little as ten dollars per treatment.

Researchers at Genentech Corporation are searching for a human hormone essential to body growth. It could be used to overcome dwarfism. Treatment of one child for this genetic defect now requires extraction of hormones from the pituitary glands of fifty human cadavers each year.

There is also talk—just talk—that the aging process might be slowed or even delayed in humans by recombining DNA in cells.

Twelve years ago, Dr. James Bonner predicted that the "capability [to live 200 years] will be within our grasps in the next several generations."[3] Dr. F. Medvedev from the Russian Institute of Medical Radiology notes that the older a person is, the more likely he is to produce a child with a genetic defect. The life span is calculated to prevent us from having children before mutations reach a scale harmful to the species. Dr. F. Marott from the Boston University School of Medicine thinks the autoimmune mechanism is switched on with the advance of aging. By the release of antibodies, we devour ourselves. Another theory is that DNA is progressively shut off as we age.

I'm inclined to think that aging is programmed into our cells and that little can be done about it. Scripture says, "It is appointed for men to die cnce, and after this comes judgment" (Heb. 9:27). That doesn't mean you must die at a specific, foreordained time. It does mean that death is certain and there is no way to avoid it, short of the return of Christ.

The average life span shot up from 47.3 years in 1900 to 68.2 years in 1950. That's an increase of almost twenty-one years. Since 1950, the average life in the U.S. has increased only about five years. Some scientists think that under optimum conditions we will go no higher than 110.

The only persons we know to have lived extremely long lives are the patriarchs of Genesis 5. These people, I think, had special genetic constitutions selected by God. They were not random samples of the human race.

Recorded human life spans shortened dramatically after the Flood, even though Abraham lived to be 175. I suspect that the wives of Noah's sons were from the short-lived varieties of humans. Their genes were mixed with the long-lived variety (Noah's sons) and life expectancy began dropping. Later the Bible speaks of seventy as the age (in round numbers) to which a man could expect to live. It's interesting that we're still hovering around that mark in this day of scientific genius.

I'm not optimistic about the DNA researchers finding a way to control aging. All of us had better be prepared to die. I do predict that many new substances will be developed through recombinant DNA research for treatment, and perhaps control in some instances, of a number of troublesome diseases. Whether science will be able to deal with these diseases at their hereditary source is another matter.

Some optimistic researchers say man will learn how to manipulate the control mechanisms in his DNA. It will then be possible, they say, to take some of a person's DNA and grow spare organs in the lab to be ready when his old ones wear out.

The real dreamers think the anatomy of man can be reshaped. One can have extra thumbs, a larger brain, longer or shorter legs, and perhaps X-ray eyes and super-hearing. It is even speculated that twenty-fifth-century man will consist of a central brain connected to a variety of indestructible limbs and sense organs that may travel as he fancies.

There are also predictions that it will one day be possible to eradicate "undesirable" attitudes and feelings. The ultimate goal is human perfection.

I'm not saying all of this will happen, nor that God will permit it to come to pass. I am saying that biologists have already accomplished many feats which were once thought impossible and that others are on the horizon.

Nor am I saying that all of this is wonderful. I have strong, mixed feelings about the potential benefits to be gained from radical and dangerous biological research. Maybe God will decide that man has become too proud and self-sufficient and will allow a one-celled monster to escape a lab and devastate a city just to bring man off his throne. The ten plagues of Exodus surely brought Pharaoh down from his pinnacle of pride and rebellion.

Some scientists speak of evolution in terms of God. They warn their colleagues against tinkering with the delicate mechanisms of evolution. Others see evolution as an enemy to be conquered. "Evolution brought us typhoid and yellow fever epidemics and outbreaks of polio," they say. "Evolution will kill us all if we don't get control of our destiny."

What can we say as Christians?

We can say that God is in control. Nothing that happens ever surprises him.

We can say that there is meaning in creation. God created life "after its kind" and divided the kinds for good reasons. He created the processes by which we are conceived, born, and ultimately die. He gave all of life a beautiful balance. Man needs to be careful about how he upsets this balance and questions the wisdom of God.

Apart from such philosophizing, we must face the real questions which recombinant research confronts us with now.

What are the permissible boundaries for experimentation with the basic substance of life?

When do the risks become greater than the potential benefits?

Who will provide the answers these questions require?

What is perfection? Is it the absence of discomfort and pain? Is it a Charles Atlas physique?

Will the absence of disease insure a perfect world? Can selfishness and hostility be genetically programmed out of man and replaced with unselfishness and love?

Who will decide which traits are worth keeping and which should be altered or eliminated?

What should be the Christian response to the challenges of science in the biorevolution?

WHERE
DO WE GO
FROM
HERE?

A professor is going around the country telling audiences, "If you're alive twenty years from now, you'll live forever."

His name is F. M. Esfandiary and he teaches courses in futurology at New York's New School for Social Research. He predicts that within two decades science will have conquered disease, pain, and even death. Within five to ten years, he says, there will be a microcomputer that can be implanted in our bodies for constant monitoring of every vital organ. If your heart starts acting up, for example, the computer will let you know in plenty of time to get expert medical attention.

The aging breakthrough, according to Professor Esfandiary, will come through genetics. Science will find a way to alter DNA so the body will keep reproducing healthy cells forever.

I know of no qualified geneticist who has made this claim. But because Esfandiary claims to be a scientist and a professor, the newspapers have picked up his prediction and many people will believe him. He's telling them what they want to hear.

We hear different stories elsewhere. In the film *Demon*

133

Seed, a scientist builds a super-smart computer, then the computer "ravishes" his wife. From this inspired union comes the little "demon," looking like a miniature knight in armor. I expect the next offering from Hollywood to be a Frankenstein bug that escapes from a biology lab and starts multiplying out of control. Or maybe the leading character will be a laboratory-bred six-legged man with steel-plated lungs and lasers shooting from his eyes.

This kind of nonsense makes many people think the bio-revolution is just a big put-on. "You don't really think they're going to clone a man?" "The test-tube baby is a fake to make money for the parents and a big name for the doctors. She was born like everybody else." "This DNA business is a lot of double-talk to confuse the bureaucrats and keep the grant money rolling into the universities where they've got nothing better to do."

The actual accomplishments in biological science are about halfway between this cynicism and the fantasies mentioned. The ultimates, which science fiction writers imagine, haven't happened yet, but present developments are still astounding. As an astonished minister remarked after hearing about some of the things in this book: "Pardon me while I go home and have nightmares."

At times all of us would like to turn back the clock on things that shock and disturb us. We can't. "Time marches on," as the narrator used to say in the old World War II newsreels. Indeed it does. There's no turning back. Massachusetts Governor Michael Dukakis observed in *Time* magazine: "Genetic manipulation to create new forms of life places biologists at a threshold similar to that which physicists reached when they split the atom. I think it fair to say that the genie is out of the bottle."[1] Added Robert Sinsheimer, chairman of Cal Tech's biology department: "Biologists have become, without wanting it, the custodians of great and terrible power. It is idle to pretend otherwise."[2]

The research is going to continue. Government funding

could end tomorrow, forcing university labs to cut staff, but the top scientists can find jobs with private corporations and foundations. The trend in employment of scientists is already moving in this direction. The warnings will keep coming from people in the know. Dr. George Wald predicts that the results of recombinant DNA work "will be essentially new organisms, self-perpetuating and hence permanent. Once created, they cannot be recalled. . . . It is all too big, and is happening too fast. . . . It presents probably the largest ethical problem that science has ever had to face. Our morality up to now has been to go ahead without restriction to learn all that we can about nature. Restructuring nature was not part of the bargain; nor was telling scientists not to venture further in certain directions. That comes hard."

Wald wants strong federal action to bring all research under control. He fears "for the future of science as we have known it, for humankind, for life on the Earth."[3]

Princeton's Dr. Ramsey warns that genetic engineering "could cause the genetic death God once promised and by His mercy withheld so that His creature, despite having sought to lay hold of godhood, might still live and perform a limited, creaturely service of life."[4]

There is great concern over what artificial insemination, test-tube babies with host mothers, cloning, and abortion of unwanted children will do to the family. Says Dr. Leon Kass: "The family is rapidly becoming the only institution in an increasingly impersonal world where each person is loved not for what he does or makes, but simply because he is. Can our humanity survive its destruction?"[5]

The bitterest arguments come over aborting defective children. Joseph Fletcher believes there is a moral obligation to abort children who do not measure up to a certain quality of life. He also thinks defective babies and the terminally ill should not be allowed to live if they do not meet a set of "positive human criteria." For Fletcher, an I.Q. of forty is "a questionable person," and below twenty, "not a person."[6] Fletcher also asserts that persons with genetically defective pedigrees do not have a right to reproduce. "Our gonads and gametes are not private possessions," he told the second National Symposium on Law and Genetics held in Boston, in 1979.

Dr. C. Everett Koop, who accepts the Bible as final authority, puts a different value on "imperfect" children. He says:

> When God spoke to Moses at the burning bush, he said, "Who makes man dumb or deaf, or seeing or blind? Is it not I, the Lord?" Whether you or I like it or not, God makes the perfect and also what we would call the imperfect. This is in His sovereign plan. I don't think I should say that God made a mistake and therefore I am going to get rid of this child.[7]

God's idea of perfection is different from man's. Modern man is a body worshiper. You need go no further than a television set or a newsstand to recognize this. God looks upon the inward person. I have no trouble at all believing that in God's sight a sweet, gentle Down's syndrome boy is more beautiful than a spoiled, mean-tempered Olympic star.

There seem to be two extreme views of science held in this country. One view deifies science as the potential savior of all mankind. The other places science on a level with demonism. Some Christians I know fall into the latter group. They automatically assume science is out to destroy their faith. Or they immediately become defensive whenever a scientist begins talking about origins. I think that deep in their hearts these Christians suspect that one can't be intellectually honest and still accept the Genesis record. Scientific creationism, however, assures us we can keep our minds in gear and still believe the Bible.

Science itself is a method of exploring and amassing knowledge of God's creation, then showing us how we can put this knowledge to good use. This is in the mandate of Genesis 1:28 where God commands man to "subdue" the earth. The earth was made for man to control and use in ways that honor God.

Most of the scholars who laid the foundations for modern astronomy, physics, and biology believed this. Most scientists today tend to ignore the faith of their forerunners. When you remind them that Isaac Newton, the "grandfather" of the rocketry that sent a man to the moon, was also an eminent theologian, they roll their eyes and say in effect, "Every great man has been a little foolish." They assume or pretend that the Bible and Christian faith had nothing to do with the scientific achievements of such great scientists as Newton, Pascal, and Galileo who were believers first and then scientists. Are you surprised to find Galileo on the list? He was a believer. He just happened to be out of step and way ahead of religious dogmatists who had forgotten that God created the "heavens" as well as the earth.

Science is one thing, philosophy another. Yet one's philosophy determines what he thinks about the values and purposes of science and how he uses science.

Let's get down to basics. You can divide scientists into theists, mechanists, and agnostics. Mechanists say that if we knew everything about man, he would be totally explainable in terms of physics and chemistry. To put it another way: Man is merely matter and that is all that matters. Theists say science can never completely explain man, for the totality of man transcends the realm of knowledge available to man. In other words: Man is much more than matter and in the long run that is what really matters.

Theists who believe that there is something extraterrestrial about man subscribe to some form of deity. Not all theists are Christians, but all Christians are theists. On the other hand, some who pretend to be Christians are mechanists at heart.

This leaves only the agnostics. They try to operate on

the assumption that man cannot know whether he is machine or soul. This puts them in a tortured dilemma. Take the decision to destroy a defective unborn child. From the mechanist perspective, the child is only an imperfect machine in the making. From the theist viewpoint, the child is something above and beyond a machine. Is that "something above" worth saving? If the scientist goes ahead and acts as a mechanist would, he will always wonder if he made a tragic mistake.

The true mechanist would not worry about the consequences of destroying a defective child. For him, there is only a difference in degree, not a difference in kind, between a defective human and an insect.

Actually, most mechanists aren't the monsters we theists sometimes make them out to be. Some agonize over decisions to abort children. Physical life is sometimes more precious to them than to believers. The flesh and this span of mortality are all they have.

The present generation of mechanists has no real hope of achieving immortality for itself despite the foolish talk by Professor Esfandiary and others that in a few years man will be living forever. If this could happen, think of the population problem and the drain on energy resources that would result! "Oh, but man will find a way to populate other planets," the dreamers say. Ridiculous. Man would have to take his environment with him. None of the other planets in our solar system can support human life.

Today's mechanists can only hope for a better life for future generations. So they talk grandly of cleaning up the mutants and eliminating disease. Yet to improve life they realize that they must destroy inferior life. How are they going to make the tough decisions about who shall live and who shall die? How will they ever agree among themselves? Scientists are not a special breed. They are no better or worse than other men. Some hold petty jealousies. Most compete for honors. Some will stab a colleague in the back. Some cheat on income tax.

Scientists and professors (actually many scientists are also teachers) have in the past been held in great esteem. Recent developments, however, show us that some researchers and academics have feet of clay.

Dr. Phin Cohen, an M.D. and biochemist, recently studied human blood chemistry under a $200,000 National Institute of Health grant at Harvard. An aide to his department chairman brought him a blank form to sign that was to be a list of grant expenditures. Dr. Cohen asked that the form be filled in first. After considerable hassle, he got the list and discovered money from his project was being used to pay people who had not worked on it. He then asked Harvard financial officials to audit all grants in his department. When the response was less than satisfactory, he went to NIH. The government audit hit Harvard for a refund of $132,000. This triggered a broader investigation. HEW auditors questioned the way Harvard had accounted for 40 percent of $37 million in federal grants and contracts to the School of Public Health. They demanded Harvard refund 7 percent to the government, approximately $2.35 million. Nobody said the money was stolen. It had been used for things not specified in the government regulations.

One result of the Harvard situation is tighter HEW rules. Universities receiving grants will have to toe the line. Where is Dr. Cohen? He was denied reappointment to the Harvard faculty and is now studying the accounting practices of one hundred other schools. He has found widespread overbilling of federal grants for medical insurance, hiding of cost overruns, and the practice of billing one project for work done on another. We knew this had been happening elsewhere, but I think most people have always thought academia was above such practices.

What about the scientists who worked for Hitler?

If it's wrong to worship scientists, it's also wrong to make them monsters of evil. I've already said that some of my brothers and sisters in Christ have been guilty of the latter. Some ministers have been guilty of unfairly lampooning scientists, especially biologists.

A minister should inform his congregation about creationism and evolutionism, mechanism and theism, and genetic engineering versus the sacredness of human life and individual rights. But unless he is well-read in the field or has had scientific training, he should let the Bible-believing scientists do most of the explaining.

There is also a place for public debates between believing and unbelieving scientists. The Institute for Creation Research has had great success in debating evolutionists on university campuses—to the extent that it's getting hard to find opponents who will go up against us.

However they do it, concerned Christians need to get moving. At this moment, the mechanists are in command and moving full speed ahead with education programs and social legislation. Occasionally, they go too far and are forced to take a step backwards. The Macmillan Company, which has a big slice of the market in public school science texts, recently decided to stop offering in its educational catalog human embryos embedded in plastic. Grisly? You bet.

We've been dissecting fetal pigs in my biology lab. If American society keeps going the way we're heading now, freshmen of the future will probably be dissecting fetal humans, perhaps "grown" for this purpose. If man is just another animal, there is no reason why this shouldn't happen.

At this moment, I'm both pessimistic and optimistic about the future. I'm pessimistic about the ominous trends in public education. Drug use, violence, and sexual immorality bother me, of course. But what concerns me more is the pervasiveness of secular humanism which recognizes no moral absolutes and sees everything as relative and conditioned by culture. The abortion of unwanted children, physically defective or not, is made a personal choice for which one must answer only to the forces within his culture. Homosexuality is a legitimate life style and demands respect. Sexuality, like abortion, is a matter of choice. Culture and environment, the relativists say, are the determinants in sex roles. There is no Supreme Deity, as well as no absolute commandments, no fixed positions in conduct.

Secular humanism is blatantly promoted in textbooks and classrooms in public and private colleges and universities and in many Christian denominational schools as well.

Here's an example from a highly regarded book used in some colleges. The excerpt is from a chapter titled "Morality and Communicational Process" by Dr. Harley C. Shands, Chairman of the Department of Psychiatry, Roosevelt Hospital, New York City. Dr. Shands is one of twenty-five authors of the study.

Under the topic, "Man, Language and Culture," Dr. Shands suggests that human behavior is reprehensible only within one's cultural group.

> All of us are, always and unavoidably, in the position of Solomon, to whom it was necessary for Nathan to say, "Thou art the man" with reference to his sin with Bathsheba, and in that of Oedipus, to whom it was necessary for Teresias to convey the same message. In both instances, the power of the consensus can be seen in the ensuing behavior of the rules: Solomon, more restrained and more routinely offending, was satisfied to repent; while Oedipus, appalled at the consensually abhorrent crime of incest, was moved to blind himself as his wife-mother Jocasta killed herself.

> These mythological examples attest to a truth daily evident in our court rooms. The defendant is not guilty until said to be guilty: "guilt" is not an absolute moral judgment, it is a relativistic consensual judgment.

141

> . . . In affairs of morality and ethical questions, we find the background of relevance that of relativity and consensus: a culture is primarily defined in its own terms. A culture is analogous to a dictionary—each word (or member) is defined in terms of other words (or members) which only have value in a system including all. This means, in turn, that *objectivity* is a myth, and perhaps a most dangerous and even lethal myth. In human affairs, there is no such thing as objectivity, and therefore *no such thing as a universal system of morality or ethics* [italics mine].[8]

Did you get this learned man's point? Nothing, *nothing* is absolute. Everything is cultural, even incest and suicide. A couple of other things should be noted. One, he classifies the biblical story with the Greek drama as myth. You see this quite often in high school literature and social science texts. Well, to some extent his biblical story is a myth, for it was David, not Solomon, who committed adultery with Bathsheba. Solomon was Bathsheba's son.

This is not an isolated case. Cultural relativism is the norm in secular higher education. This base for morality is wholly man-centered, not God-centered. Morality comes from within, not without. Everything is up to man. Jacques Monod, a philosopher guru among secular humanists and a Nobel Prize winner in 1965, calls man only a product of chance genetic mutation. Monod concludes: "Man knows at last that he is alone in the universe's unfeeling immensity, out of which he emerged only by chance. His destiny is nowhere spelled out, nor is his duty."[9]

The secular humanistic philosophy is also present in modern literature and is further expressed in television and movie dramas. These are some of the moral-ethical values which appear:

(1) The quality of personal experience holds priority over one's role as member of a family, employee and/or

employer, and citizen of the community and country. One's first obligation is to one's self.

(2) Morality is developed from within one's own mind and experiences and not from institutions—church, state, school, corporate business, which are chained to the past and are no longer relevant to present life.

(3) Moral ties and loyalties are linked only to one's immediate group.

(4) Sex is only an experience and has no moral or ethical relevance beyond self-fulfillment, which may or may not be linked with affection for another. Expressions of sexuality, whether heterosexual or homosexual, are contingent upon style and preference.

(5) Honesty is virtuous only within the intimate group. It is all right to rip off outsiders, especially uncaring institutions.

(6) There is no purpose or meaning beyond this life.[10]

No moral consensus exists today on absolute rules such as the Ten Commandments. The world view of most educators, communicators, government bureaucrats, and many denominational church functionaries can be summed up this way: Do it. Experiment. Be free. Live, live, live. You are all you have. You only go around once, so make the most of life while you have it.

This philosophy is tearing apart our social structure, breaking up marriages, and alienating members of families from one another. There is no place here for defective children or any other misfits, since they will only tie the "beautiful people" down. Of course, if caring for a handicapped child or aged grandma is your bag and you get enjoyment from it, then do it. Whatever makes *you* feel happy and fulfilled is OK.

This is the exact opposite of the Christian life style which Jesus asked his disciples to follow: giving instead of getting, serving instead of seeking self-satisfaction, putting others' interests before your own, and seeking to please and honor God more than anything else. "Whoever wishes to save his life shall lose it," Jesus said, "but whoever loses his life for My sake, he is the one who will save it" (Luke 9:24). Witness the unhappy multitudes who drag from one weary experience to another, seeking a new "high." Witness the suicides of rock stars and movie idols from drug overdoses. "Vanity of vanities! All is vanity" (Eccl. 1:2).

What does this have to do with the moral problems posed by the biorevolution? Everything. If man answers to no one but himself, then science must seek to make the perfect man. If unborn babies cannot contribute to the self-fulfillment of parents and others, then abort them, whether they happen to have physical limitations or to be of the wrong sex. If newborn babies reveal a low "quality" of life, then allow them to die by withholding treatment. Before the Supreme Court authorized abortion on demand in 1973, Dr. Koop predicted that infanticide would follow after a million abortions a year in the U.S. We've reached that figure. Dr. Koop now attests that infanticide is "being practiced widely in this country today, that is, the deliberate killing by active or passive means of a child who has been born."[11] Pediatrician Raymond Duff reported in the *New England Journal of Medicine* (October 25, 1973) that physicians at the Yale-New Haven Hospital withheld treatment from a group of sick babies born with physical defects. Fourteen percent died as a result.

This is certainly not happening in every hospital. But it is happening in some institutions of "mercy" in the U.S.

In the face of all this, what are we Christians to do?

We must begin with our children at home. This will mean regulating TV viewing and reading by our younger children. It certainly means setting the proper example and keeping the lines of communication open. As our children get older, they will inevitably be exposed to the values of the world. If they've been reared properly, they'll know where these clash with Christian principles and be able to choose responsibly.

Our son graduated from a public high school. Our sixteen-year-old daughter is continuing in a public high school. But we're committed to Christian schools all the way for our four-year-old daughter. Maybe ten years ago public schools would have been OK for her. But the climate has deteriorated so rapidly that for her protection and nurture we intend to put her under Christian educational training. I know that public school is still the only option for many Christian parents. I feel for them. I wish there was a tax voucher system so that every family could afford to send their children to the school of their choice.

We also want the best Sunday education for our children. I'm very impressed by the Sunday school at the Thomas Road Baptist Church where we now belong. Little Jennifer is learning Bible stories and Christian songs every Sunday.

I'd like to put in a plug for improved Christian education on Sunday. In most churches, there's no homework, no reciting in class, no tests, and no grades. Why call it "school"? You just fool yourself if you think that thirty minutes in a stuffy Sunday school room under an untrained teacher reinforces your children against the secular humanism they encounter in public schools, novels, music, television, and the movies.

I'm not advocating that we pull out of the educational system. I am saying we must bring our children up in the "discipline and instruction of the Lord" (Eph. 6:4). We protect them from disease and accident. Why not protect them from the poison of secular humanism as much as possible?

Nor am I advocating that we start a war against biological research. On the contrary, I propose that we encourage more Christian young people to enter scientific careers, especially teaching.

They first need a good Christian upbringing and exposure to the biblical world view. To be adequately prepared, they may need to do their undergraduate work at a Christian college where the faculty believes in the authority of the Bible. Unfortunately, some "Christian" colleges are more humanistic than they are Christian. Personally, if I had no other option, I would rather send my child to an outright secular school where humanists talk like humanists than to a religious college where humanists use the name of God to deceive their supporting constituency.

Most students will probably have to do their graduate work at a secular institution. By this time the student should be mature enough to face the conflicting philosophies that swirl in higher education. In general he will be treated fairly, although he may have an occasional run-in with an unbelieving professor or be put on the spot in a seminar. If he keeps his composure and presents his case in a pleasant, constructive manner, he will probably not be persecuted. If he is consistently treated unfairly, he can complain to the proper authorities about discrimination. Administrators are very sensitive about discrimination today. Of course, any Christian student should strive to do his best academically.

We can't afford to be negative toward scholarship. Maybe we won't educate many top-flight researchers and faculty for the most prestigious universities. But if we want our Christian colleges to be accredited, then we must have more teachers with earned doctorates.

Up to this point, I've talked about what we can do for ourselves. We also have a responsibility to our community and our country in regard to the direction in which the biorevolution is going. You're against human cloning? So am I. Cloning to improve agriculture and to improve the breed of domestic animals is valid, but cloning humans is one area of research which should be left strictly alone. I've already stated my reasons for this.

Artificial insemination for lesbians should be stopped.

At the very least, a doctor shouldn't be required to inseminate a lesbian if she demands it.

Host motherhood is a little different from artificial insemination at the request of a married couple. I think there should be restrictions against a woman bearing a baby for another. There are just too many problems involved, both biblical and psychological. I don't advise artificial insemination by donor, even though my objections against this practice are not quite as strong as those against host mothers.

Recombinant DNA research, as I discussed in the previous chapter, is a mixed bag for me. It holds great promise for curing some genetic diseases. It also has vast potential for wrong. Proper safety precautions must be observed to prevent a dangerous "bug" from escaping and contaminating people. I don't like government regulations any more than the next fellow, but somebody has to set guidelines, and I'm not in favor of the researchers policing themselves. What worries me more is the future possibilities for changing heredity. I'm not ready to make any specific pronouncements here. I'm going to continue to be concerned and watchful.

Experimentation on humans is another very serious matter. The record shows that scientists can't always be trusted to get proper consent. The possibility of research on live embryos and fetuses for the advance of science makes my blood run cold.

I'm presently opposed to amniocentesis. I recognize that some Christian professionals will differ with me here, but I must agree with Dr. Koop that amniocentesis is a "search and destroy" mission. Except in unusual situations, a woman will have this test planning to have an abortion if the results turn out wrong.

I will modify my view when new ways are found to treat unborn babies with genetic disease. To my knowledge, galactosemia is the only malady that can now be identified and treated in the womb. This is an extremely rare ailment. Its existence doesn't justify thousands of women taking the test.

In the future we may be faced with some touchy legislation about amniocentesis testing. Unless we get a constitutional amendment forbidding abortion on demand or the Supreme Court reverses the 1973 decision on abortion (which I don't think is likely), we may soon be forced to abort defective babies.

Frankly, I was pessimistic about the future of America until this past summer. I just didn't see any way the trend could be stopped. I felt that the freedom and Christian values which we cherish were disappearing forever.

I've gained new hope since coming to Liberty Baptist College. Liberty was founded by Dr. Jerry Falwell, pastor of Thomas Road Baptist Church. Dr. Falwell, as you probably know, has one of the largest and most respected Christian television ministries in the world.

Dr. Falwell believes God is not finished with America. He thinks this country has a future purpose to be fulfilled in evangelizing the world. He's convinced me that Christians can do something to turn back the tide.

He has set out to hold "I Love America" crusades in every state. He is promoting a movement called "Moral Majority" which calls for Americans of all religious persuasions to unite for a moral reformation. This will involve political action, writing letters to legislators, and supporting candidates who will vote the right way on moral issues.

This scares some people. They say we're trying to impose our moral views on everybody. Yet there's plenty of moral legislation already on the books—laws against murder, theft, and perjury, etc. We need laws to protect the unborn and to keep scientific research within moral bounds. We're not calling for an inquisition. We just want to guard public morals, protect the helpless, and save our country from being destroyed before it fulfills God's purpose. This anything-goes moral pollution must be stopped.

We Christians, of course, must not dodge our responsibilities about genetic disease. Genetics is something most of us know little about. Some of us have displayed our ignorance in the past by labeling certain diseases as "judgments" from God. Then science comes along and discovers a genetic cause. We can get into trouble by drawing inferences from Scripture that aren't there.

We need to bone up on genetics. Pastors who do premarital counseling should be aware of genetic problems and should find if any exist in the families of engaged couples. When trouble is suspected, couples should be referred to trained genetic counselors. I say this because genetic risks are not always easy to determine. There are too many exceptions to every generality, and working out the risks involves mathematical procedures which may be difficult for the untrained person.

What should Christians do upon becoming aware that they have bad genes? I can't give a pat answer here. There are some moral questions on which Scripture does not speak clearly and this is one. What about the command to "be fruitful and multiply"? That was given to Adam and Eve. They were in perfect genetic health. It couldn't apply to every future adult. Many people are sterile and can't have children.

It is true that children are a gift of the Lord. When I talk to my students about family planning, I ask: "Do you have the right to accept the gift of a child if you're not prepared to care for it? Does the potential to produce a child authorize you to go ahead when you can't feed an extra mouth?"

The same question applies to a couple who face a good chance of having a defective baby.

After our last child was born, the doctor told my wife

that the next one might kill her. I believe God through the doctor gave this information so I could make a wise decision. For this reason I had a vasectomy. My wife is more important to me than a fourth attempt at fatherhood.

Consider a couple who has found they have one chance in four of having a child with cystic fibrosis. I would tell that couple: "Maybe God doesn't want you to have a child. Maybe he wants to do something special with your childlessness." I would speak the same way to a person who is homosexual through no fault of his own (a small percentage of homosexuals are this way). I would suggest, "Maybe God wants to do something special through your celibacy."

Isn't the ability to reproduce an indication that doing so is a God-given right? Not necessarily. There are times, as I mentioned before, when we should consider not having children or at least postponing their birth. There are some couples who should probably consider sterilization.

There is also the matter of obedience. True Christians in a biblical sense have no rights. We belong to Christ. We should seek his will, not our own. Whether we have children or not should be decided in conjunction with his perfect will.

One final word. How should a Christian couple respond to the birth of a defective child? We've been talking about genetic risks, yet every couple faces the possibility of having a child that is less than normal.

I can tell you what one pastor and his wife did and how God blessed them for making the right response. The late Dr. Donald Barnhouse, a magnificent preacher, tells the story.

Dr. Barnhouse was conducting a week of services in a large church at the time when the pastor and his wife were expecting their first child. The parents-to-be were anxious all week and Dr. Barnhouse tried to lighten the tension with humor.

On the last night Dr. Barnhouse went to the platform and waited. When the pastor didn't come, Dr. Barnhouse conducted the entire service by himself.

The pastor arrived shortly before the end of the service and sat solemnly in the back of the sanctuary. When Dr. Barnhouse asked, "Is everything all right?" the preacher said, "Could I see you in my study?"

There the pastor blurted out his agony. "Our child is a mongoloid. My wife doesn't know yet. What am I going to tell her?"

"My friend, this is of the Lord," Dr. Barnhouse said solemnly. He showed the young pastor the passage from Exodus which Dr. Koop is fond of quoting today: "Who has made man's mouth? Or who makes him dumb or deaf, or seeing, or blind? Is it not I the Lord?" (Ex. 4:11).

The pastor opened his Bible to study the verse. As he looked at it, Dr. Barnhouse added, "You know the promise in Romans 8:28, that God causes all things to work together for good to those who love him. . . ."

The pastor took his Bible and went straight to the hospital. As he walked in his wife was crying, "I want to see my baby. I've asked to see my baby, and they won't let me. Is anything wrong?"

The preacher read the verses. "My precious sweetheart, God has blessed us with a mongoloid child."

The young wife and mother cried long and hard. Finally she asked to see the verse from Exodus. She read it and thought for a while.

Finally she said, "I've got to call Mother."

As Dr. Barnhouse tells the story, there was a switchboard operator in this hospital who not only heard everything that was said but was cynical about Christians. She already knew that the pastor's wife had given birth to a mongoloid child. She wanted to hear her go to pieces. So when the pastor's wife called her mother, the operator listened in.

"Mother," she heard, "the Lord has blessed us with a mongoloid child. We don't know the nature of the blessing, but we do know it's a blessing." The operator heard no sobs, no hysteria, no breakdown. She could hardly believe a Christian could react this way. When she finally accepted it, she began telling everyone in the hospital.

The following Sunday the pastor was back in his pulpit. He did not notice that in the large congregation were the telephone operator and seventy nurses from the hospital. At the conclusion of that service, he gave his customary invitation to accept Christ. "If you've never met my Lord," he said, "I want to invite you to meet him today. Come down to the altar and receive him as your personal Lord and Savior."

The pastor bowed his head. Seldom did anyone come forward in that church and he expected to give the benediction in a moment. Then he heard footsteps, heels clicking against the floor. It was like a parade. Thirty nurses came forward. Thirty nurses who had heard that the pastor's wife had said, "God has blessed us with a mongoloid child."[12]

That beautiful, true story expresses my feelings about the sovereignty of God in our lives better than anything else I could say in closing this book. What unbelievers see as a tragic mistake, God can make a blessing. I can't explain all the suffering caused by genetic diseases. I can only say that in his infinite wisdom God has a purpose in the birth of each child. It is through suffering that we are purified and taught to love in deeper ways than we would otherwise know. Sometimes the teacher which God appoints for us is a little child born with severe physical or mental handicaps.

There is a mystery here that should cause us to reflect deeply, very deeply, on some of the motivations and intended consequences of the biorevolution. Where will it lead us? Where is God's guidance? Where is the fear of the

Lord that Proverbs says is "the beginning of knowledge"? (Prov. 1:7).

Man's knowledge is a precious resource. When he uses it to protect and enhance the dignity and sanctity of human life, God will bless it. But let him tamper with God's order of life for his own ends, and man places himself in grave danger. The day of reckoning will come.

NOTES
AND
RESOURCES

CHAPTER ONE

1 David Rorvik, *In His Image,* Pocket Books edition, (Philadelphia: J. B. Lippincott Co., 1978).
2 James D. Watson, "Moving Toward the Clonal Man—Is This What We Want?" *Atlantic Monthly* (May, 1971).
3 Joseph Fletcher, "Ethical Aspects of Genetic Controls," *New England Journal of Medicine,* 285 (1971): 776-783.
4 Paul Ramsey, *Fabricated Man* (New Haven and London: Yale University Press, 1970), pp. 104, 137.

CHAPTER TWO

1 *Archives of General Psychiatry,* October, 1964.
2 cf. Isaiah 7:14; Matthew 1:18, 23; Luke 1:31-35.
3 Reported in *Science News,* February 17, 1979.
4 *The Philadelphia Evening Bulletin,* February 23, 1968.
5 *Family Weekly,* March 4, 1979.
6 *Time,* April 19, 1971, p. 51.
7 Robert T. Francoeur, *Utopian Motherhood,* (Garden City, New York: Doubleday, 1970).
8 Paul Ramsey, *Fabricated Man* (New Haven and London: Yale University Press, 1970), p. 69.
9 Emily and Per D'Aulaire, "Clones: Will There Be Carbon Copy People?", *Reader's Digest,* (March, 1979): 95-98.

CHAPTER THREE

1 I say "substantial" because there is a possibility of mutations occurring in body (somatic) cells after conception, or in this case, after cloning.
2 *Nature,* January 8, 1979.

3 James C. Hefley, *Scientists Who Believe,* (Elgin: David C. Cook Publishing Co., 1963) p. 42.

CHAPTER FOUR

1 *New York Times,* May 2, 1969.
2 Paul Ramsey, *Fabricated Man* (New Haven and London: Yale University Press, 1970), p. 71.
3 *Ibid,* p. 72.

CHAPTER FIVE

1 Donald MacKay, *The Clockwork Image,* (Downers Grove: InterVarsity Christian Press, 1974), p. 57.
2 James C. Hefley, *Life in the Balance,* (Wheaton: Victor Books, to be published in 1980).
3 *Ibid.*
4 *Ibid.*
5 James C. Hefley, *Dictionary of Illustrations,* (Grand Rapids: Zondervan Publishing House, 1971), p. 127.
6 Paul Ramsey, *Fabricated Man,* (New Haven and London: Yale University Press, 1970), p. 89.
7 *Washington Post,* September 30, 1967, as quoted in *Fabricated Man,* p. 102. Also cited in *In His Image,* p. 56.

CHAPTER SIX

1 David M. Rorvik, "The Embryo Sweepstakes," *The New York Times,* (September 15, 1974).
2 Allen R. Utke, *Bio-Babel.* (Atlanta: John Knox Press, 1978), p. 35.
3 Dietrich Bonhoeffer, *Ethics,* translated by N.H. Smith (New York: The Macmillan Co., 1955), p. 131.
4 Fletcher, Joseph, *Morals and Medicine,* (Boston: Beacon Press, 1960), pp. 150-151.
5 Howard Brody, *Ethical Decisions In Medicine,* (Boston: Little, Brown and Company, 1976), Appendix IV.
6 *Ibid,* pp. 156-58.
7 *Ibid,* p. 159.
8 Kenneth Guentert, "Will Your Grandchild be a Test-tube Baby?" *U.S. Catholic* (June, 1977).

CHAPTER SEVEN

1 Personal interview with James C. Hefley.
2 *Washington Post* (February 3, 1980).
3 Dr. Laurence E. Karp, *Genetic Engineering: Threat or Promise* (Chicago: Nelson-Hall, 1976), pp. 137, 158.
4 James and Marti Hefley, "Babies in Question," *Today's Health,* (August, 1970).
5 *Ibid.*
6 Newspaper column, "My Answer."

7 Norman Anderson, *Issues of Life and Death,* (Downers Grove: InterVarsity Press, 1974), pp. 49, 50.
8 Harman Smith, *Ethics and the New Medicine,* (Nashville: Abingdon Press, 1970), pp. 72, 73.
9 "Babies Made to Order: How to Stir a Tempest in a Test Tube," *Chicago Daily News,* (September 20, 1974).

CHAPTER EIGHT

1 Robert Cooke, *Improving on Nature,* (New York: New York Times Book Co., 1977), p. 15).
2 *Ibid.*

CHAPTER NINE

1 Susan Jacoby, "Should Parents Play God?" *McCall's,* (March, 1977).
2 "Medical Ethics and the Stewardship of Life: An Interview with Dr. C. Everett Koop," *Christianity Today,* December 15, 1978.
3 Associated Press, September 6, 1979.
4 Richard M. Restak, *Pre-meditated Man,* (New York: Viking Press, 1973).

CHAPTER TEN

1 *Time,* April 19, 1971, p. 43.
2 *Newsweek,* August 20, 1979, p. 41.
3 *Science Digest,* March, 1967, p. 52.

CHAPTER ELEVEN

1 *Time,* April 19, 1977, p. 32.
2 *Ibid.*
3 George Wald, "The Case Against Genetic Engineering," *The Sciences* (September/October, 1976).
4 Paul Ramsey, *Fabricated Man* (New Haven and London: Yale University Press, 1970), p. 96.
5 *Time,* April 19, 1971, p. 51.
6 Howard Brody, *Ethical Decisions in Medicine* (Boston: Little, Brown and Company, 1976), Appendix III.
7 Mark Fackler, "Abortion: Two Views," *His* (February, 1979): 18.
8 *Communication: Ethical and Moral Issues,* Edited by Lee Thayer, (London: Gordon and Breach Science Publishers, 1973).
9 Jacques Monod, *Chance and Necessity,* (New York: Alfred A. Knopf, 1971), p. 180.
10 These and other themes are postulated by Winston Weathers, Professor of English and Director of the Rhetoric Program at the University of Tulsa in his chapter titled "Literary Communication" in *Communication Ethics and Moral Issues.* (See note 8 above).
11 Fackler, "Abortion: Two Views," p. 18.
12 Dr. Barnhouse's story has been repeated by many ministers. This version is from a sermon by Ben Haden, Pastor of First Presbyterian Church, Chattanooga, Tennessee.